BREAKTHROUGH
Revealing the Secrets of Rebreather Scrubber Canisters

Praise for Breakthrough

"One of the most important factors in rebreather design is scrubber efficacy. It is also one of the least well understood. Generations of rebreather divers have guessed at how long their scrubbers will continue to scrub CO_2 when using their equipment in environmental conditions outside of those in which they were tested… which is 99% of the time. Other users have incorporated personal use practices based on untested community lore and myths. John's book dispels many of these unsafe practices and provides a base for making informed decisions regarding real-world diving practices vital for the rebreather diver. It is a "must read" for all serious users of this equipment." Jeffrey Bozanic, PhD, Author of Mastering Rebreathers and Senior Technical Editor of the NOAA Diving Manual.

"BREAKTHROUGH fills the knowledge gap for mathematical and computer modeling of carbon dioxide (CO_2) elimination in underwater breathing apparatus (UBA). A must-read for every UBA designer or diver interested in a better understanding of scrubber canisters and metabolic CO_2 elimination." Vincent Ferris, Assistant Program Manager for Department of the Navy

"John Clarke's new monograph "Breakthrough: Revealing the Secrets of Rebreather Scrubber Canisters" is truly that! The retired scientific director of the US Navy's Experimental Diving Unit (NEDU) is the first to illuminate the inner workings of what has literally and figuratively, been a "black box"—the carbon dioxide (CO_2) scrubber canister, which is a critical component of every rebreather—based on decades of existing and new research conducted by the author and the US Navy.

To accomplish this feat, Clarke derives a set of deterministic, stochastic, and visual models that enable us to virtually see and understand the underlying physical processes carried out by a CO_2 scrubber and how they are affected by the diver. With the help of these tools, he is able to answer important, long-standing, and practical questions surrounding scrubber use and the impact of numerous factors on scrubber efficacy and duration, including temperature; scrubber packing and insulation, absorbent granule size and distribution, the differences of axial or radial scrubbers, as well as a host of biological variables.

Throughout the exposition, Clarke, a respiratory physiologist, presents the highly technical material in a fascinating, easy-to-absorb manner that makes it accessible to motivated readers. As such, Clarke's "Breakthrough" should be required for every rebreather diver." Michael Menduno/M2, Editor-in-Chief, InDEPTH.blog

Diving Novels by the Author

The Jason Parker Trilogy:

Middle Waters

Triangle

Atmosphere

BREAKTHROUGH
REVEALING THE SECRETS OF REBREATHER SCRUBBER CANISTERS

JOHN R. CLARKE, PH.D.

Wet Street Press
Panama City

Published in the United States of America by Wet Street Press, Panama City, FL.

Breakthrough: Revealing the Secrets of Rebreather Scrubber Canisters – Copyright © 2022 by Wet Street Press. All rights reserved. No part of this book may be reproduced, scanned, stored in a retrieval system, or transmitted in any form or by any means—electronic, photocopying, Internet, recording, or otherwise—without written permission from the author. Please do not participate in or encourage piracy of copyrighted materials in violation of the author's rights.

ISBN-13: 978-0-9863749-7-5

ISBN-10: 0-9863749-7-0

Library of Congress Control Number: 2022947828

Cover Design by John Clarke
Cover Photo by Bernie Campoli

Disclaimer: The views expressed in this publication are those of the author and do not necessarily reflect the official policy or position of the Department of Defense or the U.S. government. The public release clearance of this publication by the Department of Defense does not imply Department of Defense endorsement or factual accuracy of the material.

Permission granted by Wolfram Research to quote Steven Wolfram does not imply an endorsement by Wolfram Research.

DEDICATION: THE UNSUNG HEROES

This book would be impossible except for the U.S. Navy men and women, military and civilian, who explore and record how underwater breathing apparatus (UBA) respond when taken to its limits. Of note, they recently risked their health by testing and evaluating UBA during the COVID-19 Pandemic while others huddled behind closed doors at home. It is not a glamourous job, but it is *mission essential*.

Some unsung heroes had previously risked their lives on undersea national security missions. They know the meaning of trusting their lives to experimental UBA. In fact, their quiet heroism garnered them starring roles in the fictional diving series, *The Jason Parker Trilogy*.

Two of the women in the novels were based on exceptional women engineers who have led the Test and Evaluation department of the Navy Experimental Diving Unit (NEDU).

So, here's to the men and women of Navy diving and NEDU T&E.

To those who do not dive, the danger and beauty of the undersea world would be nothing but an abstraction were it not for photographers who trust their UBA to keep them alive while they are focusing on their subject matter. The greatest of Navy photographers, in many people's opinions, is Bernie Campoli. His photo, taken in Morrison Springs, Florida, graces the cover of this book. In an act of raw courage, Campoli once filmed a nuclear submarine as it drove past him, dangerously close. His vulnerable body was invisible to the submarine's skipper, a vessel weighing tens of thousands of tons.

Campoli's first published photo appeared in Skin Diver magazine in the mid-1950s. According to Campoli, his work has been used on Discovery, ABC, the Learning Channel, History and Military channels, as well as motion pictures.

As of the publication of this book, Campoli remains active in the Academy of Underwater Arts and Sciences.

ACKNOWLEDGMENTS

I am forever indebted to CAPT George Bond, USN (ret.) who trained me on Navy diving physiology, and taught me about saturation diving. He was one of the GREAT ones. Also, Marc J. R. Jaeger, M.D. who introduced me to deep saturation diving research at NEDU (450 msw, 1500 fsw); former SEALAB Diving Medical Officer, CAPT Mark E. Bradley, MD, who hired me at the Naval Medical Research Institute (NMRI); and both CAPT Edward T. Flynn, M.D. and CAPT Edward D. Thalmann, M.D. who were the best bosses at NMRI any biomedical scientist could hope for.

As an author in a field as technical as rebreather diving, I am deeply indebted to knowledgeable reviewers. Below are a few.

Vincent H. Ferris: Vince contributed some of this book's statistical rigor and terminology. He is a program analyst for the Department of the Navy interested in diving life support system RDT&E. He has been a cave diving instructor for the National Association for Cave Diving and a mixed-gas guide to the world-renowned deep cave systems of Indian Springs, Eagle's Nest, and Diepolder Sinks. Over decades of open-circuit and closed-circuit diving, he has explored WWII Atlantic Coast shipwrecks and submerged and dry caves in the U.S., Mexico, and the Bahamas.

Michael Menduno: Michael Menduno/M2 is the editor-in-chief of Global Underwater Explorer's *InDepth* online magazine, a Senior Editor with DAN Europe, and an award-winning journalist and technologist who has written about diving and diving technology for more than 30 years.

Jeffrey Bozanic: Ph.D., author, technical diving, and rebreather instructor. Jeff has been diving rebreathers since 1988, amassing over 2,500 hours on a wide variety of units. Recipient of the DAN Diver of the Year Award, the American Academy of Underwater Sciences (AAUS) Conrad Limbaugh Award for Scientific Diving Leadership, and the Beneath the Sea Diver of the Year Award (Science). He heavily edited this manuscript, provided many Antarctic diving photos, and generously answered questions about his team's recent rebreather operations under Antarctic ice.

John Heine: Ph.D., for sharing his pioneering work on the use of rebreathers in Antarctica. Canister durations in 28°F seawater were handled conservatively, without incident.

Donald D. Joye: Ph.D., Emeritus Professor of Chemical Engineering at Vilanova University, introduced me to the Ergun equation for pressure drop in packed cylinders.

Michael Clarke: Ph.D., Fellow of the Royal Society of Chemistry, for disentangling the chemistry of CO_2 absorption reactions by sodalime.

In an academic setting, I have both encouraged and critiqued the work of two engineering post-graduate students, Shona Cunningham, now Ph.D., and Joerg Hess. They have graciously allowed me to feature their scrubber canister modeling work in Chapter 7.

Author Photo, Holly Gardner Photography
NEDU Photos, Stephen Frink, copyrighted
Antarctic photo of divers under the ice: Martin D.J. Sayer, Ph.D. Tritonia Scientific, Ltd.
Other under ice photos by Jeffrey Bozanic.
Additional Antarctic photos and graphs: Mike Lucibella, National Science Foundation, and John Heine of the Jacksonville University Marine Science Institute.
Cover Photo, Bernie Campoli, divers Thomas Swanick and Rockie Heikkinen

PREFACE

This is a technical publication containing material by the author and others, which is difficult to access elsewhere. For the curious reader, most of the material referenced in this document can be found scattered throughout various military and academic publications. This is the first time it has been compiled into a single document accessible to the public. There is also a great deal of new material not available anywhere else.

If you are a rebreather diver or are considering becoming a rebreather diver, you are a *technical diver*. So, I do not apologize for presenting technical information. Presumably, that is what you are looking for. After all, knowledge makes for safer diving!

If you are a rebreather designer, the information contained herein may benefit your projects.

Regardless of which type of rebreather you dive, closed-circuit oxygen rebreather, semiclosed circuit rebreather, fully closed-circuit electronic, or computer-controlled rebreather, they all have a *scrubber* canister that chemically removes carbon dioxide from your recirculated breath.

This monograph is dedicated solely to scrubber canisters. In my opinion, more than one rebreather diver has lost their life due to a lack of understanding of 1) how a scrubber canister works and 2) the scrubber's operational status during a dive.

The enclosed material is not a dive planner. **Do not use it as a dive planner.** It is educational, with the sole intent to illustrate the multitude of factors, and their interactions, that influence canister durations.

Because of the specialized topic of this monograph, I recommend you become familiar with the more encompassing sources of rebreather information. For Semi-Closed Circuit Rebreathers, there is *An Introduction to Semi-Closed Circuit Rebreathers*, written by Joe Odum of Technical Diving International (1995, 2004.) For fully closed rebreathers, a sampling of the books available is: Jeffrey E. Bozanic's two books, *Understanding Rebreathers* (2002) and *Mastering Rebreathers* 2nd Edition (2010), *The Basics of Rebreather Diving: Beyond SCUBA to Explore the Underwater World (Jill's Guides)* by Jill Heinerth, and *The Simple Guide to Rebreather Diving* by Steve Barsky, Mark Thurlow, and Mike Ward, Best Publishing.

Photo credit: Jeffrey Bozanic. Titan Rebreather.

Table of Contents

DEDICATION: THE UNSUNG HEROES iii
ACKNOWLEDGMENTS iv
PREFACE vi

Chapter 1. Introduction 1

A Little Sodalime History—the U.S. Navy Perspective 1
To That End 6
Carbon Dioxide Absorption 8
Size Characteristics of Granular Sodalime 12

Chapter 2. Background 15

Unmanned Canister Duration Tests 16
A Military Oxygen Rebreather 18
Residence Time 19

Chapter 3. Prior Research 21

U.S Navy, Washington D.C. 21
Institute of Naval Medicine, Yugoslavia 23
University of Florida 25
NEDU 26

Chapter 4. Mathematical Models 27

Approach 27
Specifics 28
Derivation of Breakthrough Equation 30
Variables 35
Unmanned Testing Variation 37
Physiological Factor Variation 38
Human Data Distributions 40
    Absorbent Mass 40
    Oxygen Consumption 41

Respiratory Exchange Ratio...42
Ventilatory Equivalent 42
Sampling Error 43
Propagation of Error 45
Monte Carlo Analysis.. 52
Conclusion.. 54

CHAPTER 5. COMPUTER MODELS...57

Synchronicity... 57
Stochastic Simulated Physical Model 59
Diving Application 59
Chemistry 67
Decoupling Work and Ventilation... 75
Porosity... 84
Canister Insulation... 85
Simulation Scaling... 92
    Nanoscale variability...93
    Macroscale variability...97
Reaction Initiation 99
Water Temperature 99
Radial Canisters... 100
Flow Rate Dependency.. 102
Real Data 103
Carbonate Deposition 105
The Power of Simulation... 110
Linear Flow 111
Simulating the Impossible 112
Cellular Automata 115
Summary... 117

CHAPTER 6. APPLICATIONS 119

Cold-Soaked Canisters 119

Thermal Conductivity 122
Absorbent Granule Size Distributions 124
Friability 128
Absorbent Lot Variability 129
Specialized Canisters 131
Predive Decision Making 133
Resistance Matters 138
Resistance Limits 142
Interactions 147
Down the Rabbit Hole 149
Planned Canister Durations 152
Physiological Variation 155

CHAPTER 7. CFD 159

Ansys CFX 13.0 159
FlexPDE 161
CFD Contribution to the U.S. Navy 162

CHAPTER 8. CONCLUSIONS 165

Summary of Relevant Factors 165

REFERENCES 171
APPENDIX A: COLOR MAPPING 1
APPENDIX B: ERGUN EQUATION 1
APPENDIX C: "WORK OF BREATHING" AND FLOW RESISTANCE 1

Without Integral Calculus 4
Rebreather Corrections 5
Remember This 7

APPENDIX D: NUCKOLS, PURER AND DEASON 1
APPENDIX E: A HYPERBOLIC CASE 1
APPENDIX F: CLUSTER INDEX 1

APPENDIX G: SUPPLEMENTARY IMAGES 1
INDEX 1
ABOUT THE AUTHOR 1

LIST OF TABLES

Table 1. Scrubber canister simulation 34
Table 2. Unmanned coefficients of variation (COV). 37
Table 3. Calculated time to reach 2% inspired CO_2 49
Table 4. Percentage probability of elevated inspired CO_2 50
Table 5. Mean canister breakthrough at 2.0% CO_2 = 285 min.. 52
Table 6. Simulation results at breakthrough in 34°F water. 95
Table 7. Simulation results at breakthrough in 70°F water. 95
Table 8. NATO allowable particle distribution. 124
Table 9. Mesh size comparisons. 125
Table 10. Resistive Effort data from rebreather testing. 150
Table 11. Converting 300 fsw experimental RE data. 151

Figure Legends

Figure 1. SDV divers on MK 16 rebreathers.
Figure 2. Polar diving under the ice.
Figure 3. Typical rebreather schematic.
Figure 4. Pail of Molecular Products Sofnolime®.
Figure 5. Sodalime CO_2 absorbent in granular and rolled forms.
Figure 6. A rebreather scrubber canister filled with loose sodalime granules.
Figure 7. EX-19, the U.S. Navy's first digital rebreather.
Figure 8. Confidence and prediction intervals for Sofnolime 812.
Figure 9. Log-Normal Fit to Sofnolime 408 granule size distribution.
Figure 10. Inspired CO_2 and canister breakthrough in combat swims.
Figure 11. NEDU's Unmanned Testing Laboratory and chambers.
Figure 12. A Warrant Officer Navy SEAL with a U.S. Navy Oxygen rebreather.
Figure 13. The backside of a LAR V (MK 25).
Figure 14. The calculated effect on gas flow waveforms through a canister in a two-bag rebreather versus a single-bag rebreather.
Figure 15. Studies of the effect of ventilation parameters on CO_2 scrubbers.
Figure 16. The U.S. Navy's deepest heliox dive at NEDU. From Faceplate,
Figure 17. Four "identical" canister duration runs in 70°F water.
Figure 18. A fifth canister broke through early due to water intrusion.
Figure 19. CO_2 breakthrough curve from equation (5).
Figure 20. Dwyer and Pilmanis.
Figure 21. Morrison's correlation between R and oxygen consumption.
Figure 22. Mass of absorbent in an oxygen rebreather canister.
Figure 23. Variation of oxygen consumption, (Knafelc, 1989.)
Figure 24. Variation of the respiratory exchange ratio, R.
Figure 25. Variation of ventilation equivalent for oxygen, KO_2.
Figure 26. The relation between average canister durations and 95% prediction estimates.
Figure 27. Breakthrough curves assuming a small covariance.
Figure 28. Breakthrough curves assuming a large covariance.
Figure 29. If dived to the anticipated breakthrough time, dangerously high inspired CO_2 could be encountered.
Figure 30. CO_2 curves with absorbent mass chosen randomly.
Figure 31. Combined variance in mass and $\dot{V}_{O_2}$.
Figure 32. Combined variance in mass, $\dot{V}_{O_2}$, R and KO_2.
Figure 33. Heat flowed from hot to cold in the solid and gas phases.

Figure 34. Two sizes of modeled spherical sodalime granules.
Figure 35. CO_2 molecules encounter the granule's outer shell and diffuse.
Figure 36. Canister Options
Figure 37. Simulation Results
Figure 38. A typical rebreather gas circuit.
Figure 39. Data gathered on an AP Diving Inspiration rebreather.
Figure 40. Snapshots for a time sequence of absorption.
Figure 41. Arrhenius equation.
Figure 42. The assumed absorption probability in the SPM model.
Figure 43. A warm CO_2 absorbent canister in cold water.
Figure 44. Heat is being lost to the water.
Figure 45. The heat of reaction is carried downstream.
Figure 46. Halfway expended.
Figure 47. Color remapped to accentuate the stochastic nature of heat bubbling.
Figure 48. Fluctuations in CO_2 overflow counts per compute cycle.
Figure 49. Plots of two specific cell temperatures near the entrance to the radial canister, and average gas and granule temperatures.
Figure 50. Simulation within a section of cylindrical axial canister.
Figure 51. Physiological experiments at depth.
Figure 52. Average temperature over time at various CO_2 injection rates.
Figure 53. Breakthrough at 40°F as a function of CO_2 production rate and residence time (Tr).
Figure 54. Breakthrough as a function of CO_2 production rate and temperature.
Figure 55. At breakthrough. $\dot{V}_{CO_2} = 0.015$.
Figure 56. At breakthrough. $\dot{V}_{CO_2} = 0.1$.
Figure 57. At breakthrough. $\dot{V}_{CO_2} = 2.0$.
Figure 58. Edge and internal (colored) voids created probabilistically.
Figure 59. Effect of filling probability and edge channeling on canister breakthrough.
Figure 60. Thermal mapping of cylindrical canister cross-sections without an insulating canister shell.
Figure 61. A color map designed to differentiate between the hottest regions.
Figure 62. Insulating material surrounds the scrubber.
Figure 63. Insulation compared to no insulation during passive cooling.
Figure 64. Uninsulated canister temperature profiles.
Figure 65. With insulated cylinder walls.
Figure 66. Temperature profiles after 450 computational cycles.

Figure 67. Canister duration as a function of water temperature for insulated and non-insulated canister.
Figure 68. Definition of simulation breakthrough.
Figure 69. Means and standard deviation for five SPM runs.
Figure 70. Using the SPM to test the effect of macroscale variability.
Figure 71. The inlet end of a simulated cylindrical axial scrubber canister.
Figure 72. An example of the effect of water temperature.
Figure 73. One type of radial flow canister.
Figure 74. An exhaled breath flow path for a radial canister.
Figure 75. The effect of gas residence time on CO_2 breakthrough.
Figure 76. SPM canister durations under resting conditions.
Figure 77. Carbonate in 70°F water at breakthrough. 2mm granules.
Figure 78. Carbonate at breakthrough in 70°F water. 4mm granules.
Figure 79. Breakthrough, 2 mm granules in 34°F water. No insulation.
Figure 80. 34°F water and 4 mm granules at breakthrough.
Figure 81. After breakthrough carbonate deposition, 2 mm granules.
Figure 82. After breakthrough carbonate deposition, 4 mm granules.
Figure 83. Linear flow in Micropore ExtendAir Cartridges.
Figure 84. Minimal gas flow from left to right.
Figure 85. Adding the effect of convection from the gas flow.
Figure 86. Red circles surround white dots representing CO_2 molecules.
Figure 87. Laboratory test of CO_2 absorption in a cold canister.
Figure 88. Transient scrubber recovery before complete failure.
Figure 89. The effect of granule size on premature breakthrough.
Figure 90. A cold-soaked canister was not able to sustain its reaction intensity.
Figure 91. SPM sequence for the pre-chilled canister with lower granule conductivity.
Figure 92. Ro-Tap Sieve Shaker.
Figure 93. MeshFit analysis of a fine grain Sofnolime sample.
Figure 94. MeshFit analysis of an H.P Sodasorb sample that failed.
Figure 95. Friability test results of a Sodasorb sample.
Figure 96. The measured and expected distribution of Sofnolime 408 granule sizes match.
Figure 97. The measured and expected distribution of Sofnolime 408 granule sizes did not match.
Figure 98. The effect of asymmetrical convection on thermal distributions.
Figure 99. Leon Scamahorn and Becky Kagan Schott with ISI Megalodon Rebreathers.

Figure 100. Illustration of breathing through pores in a scrubber canister.
Figure 101. Predicted pressure drop across an absorbent bed.
Figure 102. Beds of uniformly sized granules have higher porosity than beds with a distribution of granule sizes.
Figure 103. Normal distribution of granule sizes for a mean diameter of 1.75 mm and a standard deviation of 0.4 mm, is appropriate for Sofnolime 812.
Figure 104. Activity vs σ (S.D.).
Figure 105. Boy drinking a milkshake.
Figure 106. Dr. Jere Mead, Harvard.
Figure 107. U.S. Navy meta-analysis of multiple Navy dives.
Figure 108. The calculated sum of pulmonary and canister resistance.
Figure 109. Resistance at 200, 300, 400, 600, 800, 1000 fsw.
Figure 110. The physiological loop and sources of untoward events in diving.
Figure 111. Canister duration recommendation from a manufacturer.
Figure 112. Mean, confidence limits on the mean, and prediction limits.
Figure 113. Christian McDonald and Steve Rupp prepping for an under-ice dive in McMurdo Sound, Antarctica.
Figure 114. An orange dive hut was placed over a diving hole. Photo by the author.
Figure 115. Canister temperature recordings in a Legacy Megalodon rebreather.
Figure 116. First Antarctic Dive Program CCR dive.
Figure 117. Photo by the author.
Figure 118. Axial and radial scrubber experimental set-up.
Figure 119. Test rig set-up. (a) axial scrubber and (b) schematic of thermocouple positions in the axial scrubber.
Figure 120. Cunningham's Transient Computational Fluid Dynamic model.
Figure 121. Experimentally determined CO_2 injection rate comparisons.
Figure 122. Heat, flow, and fraction of CO_2 in the canister eluent.
Figure 123. A Navy diver entering NEDU's Ocean Simulation Facility.
Figure 124. Boring a diving hole in the Ross Ice Shelf, Antarctica.
Figure 125. Every rebreather component visible adds to breathing resistance.
Figure 126. Finally, my turn.

GLOSSARY

$\dot{V}_{CO_2}$ – the volume of metabolic CO_2 produced by the human body per minute (under standard STPD) conditions.

$\dot{V}_E$ – ventilation, also RMV (Respiratory Minute Volume)

Airway resistance – the change in transpulmonary pressure needed to produce a unit flow of gas through the lungs. The pressure difference between the mouth and alveoli of the lung divided by airflow.

ata – atmosphere absolute pressure. Sea level standard pressure of 760 mm Hg is 1 ata.

Break through (verb) – CO_2 outflow from a scrubber canister reaches a pre-determined partial pressure.

Breakthrough (noun) – the name attached to the moment that canister CO_2 outflow exceeds a predetermined partial pressure.

Calcium dihydroxide ($Ca(OH)_2$ – also called calcium hydroxide. The primary material in a CO_2 scrubber canister reacts with CO_2 to form calcium bicarbonate.

Closed circuit – recirculates breathing gas without exhausting gas to the water.

CO_2 – carbon dioxide, the molecule exhaled by animals after consuming oxygen.

Discrete – described as individual items as opposed to grouped items

Drägersorb – a brand of sodalime CO_2 absorbent made by Dräger Germany.

e – the exponential function, used here to describe exponential, *ever-climbing* growth

Elastance – a measure of the ability to resist deformation when pressure is applied. The reciprocal of compliance.

Exothermic – a chemical reaction that releases energy in the form of heat.

Feret diameter – the diameter of a circle that yields the same number of pixels as the actual number of pixels for each granule image

Flow resistance – a measure of the pressure difference required to cause a unit increase in volumetric flow.

Frequency distributions – a description of the common occurrence of groups of items

Gaussian (normal) – a symmetrical bell-shaped frequency distribution

Inertance – a measure of the pressure difference in a fluid required to accelerate a mass; gas, liquid, tissue.

Kinetics – the science of movement

LAR V – an oxygen closed circuit rebreather without electronics. Also called MK 25 by the U.S. Navy.

Log-Normal – a "normal," Gaussian distribution skewed to the right.

Mean – a statistical term for "average"

Metabolic – of or produced by energetic processes within the body.

Meta-analysis – a specific statistical strategy for assembling the results of several studies into a single estimate.

Multiphysics – the coupled processes or systems involving more than one simultaneously occurring physical field and the studies of and knowledge about these processes and systems.

NEDU – Navy Experimental Diving Unit, Panama City, FL, USA

Newton's notation – various levels of mathematical differentiation symbolized by the number of dots over the basic parameter. For volume, one dot = the first derivative, flow. Two dots – second derivative, acceleration.

NMRI – Naval Medical Research Institute

Oxygen consumption – oxygen consumed metabolically. A physiological measure of work.

Probabilistic – governed by probability or chance.

Propagation of Error – engineering law that error accumulates throughout all components of a system.

R – or RER, the Respiratory Exchange Ratio

Rebreather – a type of SCUBA that recirculates breathing gas, removing exhaled CO_2 and adding oxygen as needed.

RER – or R, the Respiratory Exchange Ratio between CO_2 metabolic production and oxygen consumption.

Resistance – the change in transpulmonary pressure needed to produce a unit flow of gas through the lungs. The pressure difference between the mouth and alveoli of the lung divided by airflow.

Resistive Effort – the sensory perception of applied physical pressure, exertion in forcing gas flow. A psychophysical manifestation of volume-averaged pressure.

Residence Time – the amount of time that CO_2-containing gas remains in contact with CO_2 absorbing granules in a scrubber canister. A function of diver ventilation rate, breathing bag arrangements, canister geometry and packed granule porosities.

RMV – Respiratory Minute Volume

SCUBA – Self Contained Underwater Breathing Apparatus

Semiclosed circuit – a nonelectronic rebreather that emits gas from the loop intermittently.

SEV – Surface Equivalent Value, a way of measuring gas concentration relative to 1 atmosphere pressure

SigmaPlot – professional graphing software

Slicer Dicer – visualization software for three-dimensional data

Sodalime – generic term for granular CO_2 absorbent

Sodasorb – a defunct U.S Brand of sodalime. Made by W.R. Grace. Company purchased by Molecular Products.

Sodium hydroxide (NaOH) – Also known as lye and caustic soda, is an inorganic compound with the formula NaOH. It is a white solid ionic compound consisting of sodium cations Na^+ and hydroxide anions OH^-.

Sofnolime – a U.K brand of granular sodasorb

Standard deviation, S.D – A statistic that measures the dispersion of a dataset relative to its mean and is calculated as the square root of the variance.

Stochastic – randomly determined; having a random probability distribution or pattern that may be analyzed statistically but may not be predicted precisely.

STPD – standard scientific temperature (0°C) and pressure (1 ata), dry

TableCurve 2D – software for curve fitting data

Thermokinetics – the movement of heat during chemical reactions, especially with gas flow.

$\dot{V}O_2$ – oxygen consumption

Volume-averaged pressure – the pressure measured at the mouth averaged over a tidal volume change. Sometimes erroneously called "work" of breathing.

Chapter 1. Introduction

A Little Sodalime History—the U.S. Navy Perspective

Thirty years ago, in August 1992, a U.S. Navy diver reported that the gas in their rebreather "smelled like cat piss."

That was not a solitary complaint. It became a pervasive problem, resulting in the U.S. Navy temporarily shutting down all rebreather diving until the source of contamination had been determined.

That contamination problem occurred within a year of my move from one Navy research diving facility in Washington, D.C., to another research diving facility in Panama City, Florida.

The Navy's sodalime carbon dioxide absorbent source had become contaminated due to manufacturing changes at W. R. Grace, the makers of Sodasorb. A friend I'd left behind at the Naval Medical Research Institute in Bethesda, Maryland, Dr. Richard S. Lillo and his team identified a large amount of ammonia in samples taken from subject pails. The smaller but more toxic concentrations of ethyl and diethyl amines[1] and other hydrocarbons were even more worrisome.

My job, and NEDU's job, was to quickly qualify an alternative absorbent, Molecular Product's Sofnolime, without indicator dye. That required intense, around-the-clock testing of rebreather canister durations in every rebreather in the Navy inventory. That included the electronic mixed-gas MK 16 (Figure 1) and the nonelectronic oxygen rebreather, the MK 25, also known as the Dräger LAR V. It also included umbilical-fed saturation systems employing rebreather systems. Until we accomplished that qualification effort, there would be no rebreather diving in the Navy.

Figure 1. SDV divers on MK 16 rebreathers. (Navy photo.)

As NEDU's Testing Director of the Test and Evaluation Department, it fell to me to assure a sound scientific and engineering approach to that qualification effort. Furthermore, as Scientific Director of the overall research missions at NEDU, I had already been engaged in rewriting the Navy's Unmanned Testing Manual, Middleton and Thalmann's 1991 *Standardized NEDU Unmanned UBA Test Procedures and Performance Goals,* NEDU TR 03-81[2].

The stars were aligned, so to speak. On the one hand, we were revisiting best practices for improving canister duration testing and sodalime qualification. At the same time, we were codifying those changes in a new Unmanned Testing Manual, NEDU TM 01-94[3].

Lastly, NEDU and other NATO diving partners (Canada and the U.K.) drafted NATO Underwater Diving Standard Agreements (STANAGs 1410[4] and 1411[5])[a], which included the new U.S Navy testing procedures for UBA canister testing and sodalime qualification.

TM 1-94 was updated in 2015 by NEDU Technical Manual[6] 15-01. However, between 1994 and 2015, I found a dearth of information about the life-sustaining black box we call the "scrubber" and the CO_2 absorbent it contains. From a professional diver's standpoint, that was unacceptable.

After the new sodalime was approved by the Navy, there were numerous instances where absorbent, colloquially called "sorb," was causing problems. For example, sodalime mislabeling problems arose in the fleet and at NEDU. When the manufacturer was alerted, they said, "That can't happen." But then we sent them photos to prove it had happened.

In an especially troubling event, an alteration in a manufacturing process overseas resulted in the bizarre formation of "extruded chemical worms" interspersed among regular sodalime granules. Once again, a different manufacturer told the Navy that could not happen. Fortunately, they had no choice but to recant once we sent the photographic evidence.

Seeking even more sodalime sources, our NEDU laboratory tested Drägersorb absorbent, and I wrote the paper qualifying it for Navy Special Forces divers. At the Navy's request, we also

---

[a] *STANAG 1410 has been replaced by ADivP-05, Standard Unmanned Test Procedures and Acceptance Criteria for Underwater Breathing Apparatus, Edition A Version 1, February 2015.*

performed iterative testing of the Micropore rolled canisters, encouraging the company with steadily improving test results until they finally got it right.

After considering the consistent hemispherical shape of Dräger DiveSorb granules, I wondered how granule shape and size distributions might affect both breathing resistance and carbon dioxide (CO_2) absorption performance.

Before long, a new scrubber/sorb problem arose. We received reports that after used scrubbers had been stored in cold conditions, they would fail when returned to the water after an initial burst of scrubbing activity. Some innovative NEDU laboratory testing was able to reproduce that event.

About that time, I realized the Navy seemed to be enduring a constant stream of scrubber and sodalime events, some of which could have been life-threatening. Sadly, we had little understanding of what was happening inside those "black boxes," which were critical components of <u>every</u> underwater rebreather.

Those canister events occurred at about the same time a Navy diving officer nearly lost consciousness while pool-testing a new semiclosed rebreather.

We had no idea what was causing a steady loss of oxygen partial pressure in some, but not all, divers. So, I set about creating a computer simulation of the new style of semiclosed UBA.

What I discovered was shocking[7]. The manufacturer had inadvertently set up some of our divers for failure simply because divers were "adjusting" breathing bags for <u>comfort</u>. But that prevented the automatic nitrox demand add valve from activating. A human factors issue could have killed this diver had he not been monitored.

The UBA simulation was initially called "Semiclosed," but at the request of a Navy sponsor, it was later expanded to cover the functioning of most all Navy SCUBA. It is now called *UBASim*.

Flushed with that programming success, I approached the canister-sorb problem with the same computer simulation methods and fervor[8]. When the simulation exactly reproduced a cold-temperature-induced absorbent failure we found in the laboratory (Chapter 6), I knew we had a valid model that might be relevant to the next frontier, rebreather diving in polar regions.

Figure 2. Polar diving under the ice. Photo credit Martin D. J. Sayer.

That simulation effort was most interesting because it did not follow mathematical rules. Instead, it followed elementary physical rules familiar to most high school physics students. It was a *stochastic* approach, bounded by the usual laws of probability rather than a deterministic modeling approach.

Years after my initial discoveries from the simulation, improvements in the code are still occurring. That means more and more findings are being revealed about the goings-on inside rebreather diver scrubber canisters. Thanks to commercial software allowing the visualization of large three-dimensional data sets

(Slicer Dicer® by PIXOTEC, LLC), we can now "see" what is happening inside rebreather scrubbers.

That canister modeling effort was an extracurricular activity not paid for by the Navy. In other words, I did the coding, testing, and analysis on my own time at home. When the software was first created, home computers were relatively slow. A simulation containing just a fraction of the current "nodes" required twelve to 24 hours of simulation time to produce one 3-dimensional array of data.

In the 2020s, a simulation run takes less than half an hour with a relatively modest home computer and a much larger simulation model. Therefore, it is time to release the information this effort has revealed to all rebreather divers. That is the intent of this monograph.[b]

To That End

What if you could see inside the rebreather carbon dioxide scrubber keeping you alive underwater? Thanks to math and computer simulations, you can…in a way.

While this author and diver spent decades as the Scientific Director of the Navy Experimental Diving Unit (NEDU) and a researcher at the Naval Medical Research Institute (NMRI), I encountered few things as fascinating and mysterious as the workings of a type of underwater breathing apparatus called *rebreathers*. By *rebreather,* I speak of breathing devices far more complicated than open circuit SCUBA.

---

[b] *By 2010 the growing-pains and problems with sodalime had abated almost completely. Nevertheless, the savvy diver must remain vigilant.*

Much of the material has been archived in hard-to-find technical reports and military presentations. This is the first time that information has been compiled for the general public. In addition, new material has been created specifically for the technical diver audience. In so doing, I have <u>tried</u> to avoid the cumbersome technical jargon so enamored by scientists and engineers. If you want to get into the "nitty-gritty" of the subject, the remaining math is necessary. However, casual readers should feel free to gloss over the more difficult parts. Your comprehension won't be significantly affected.

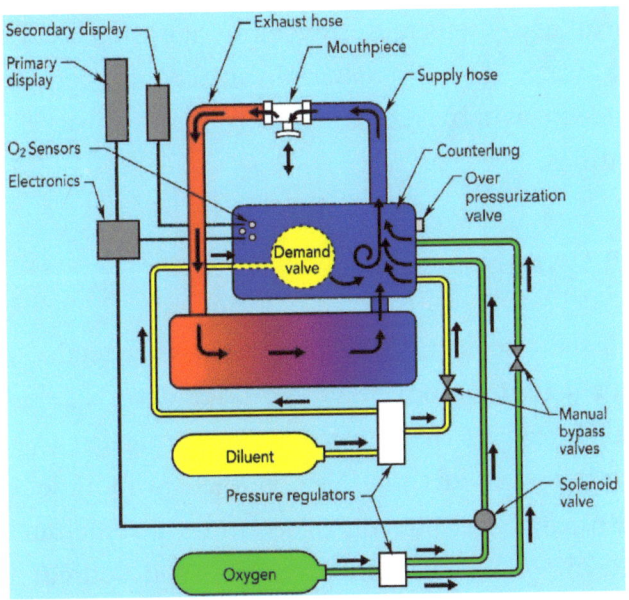

Figure 3. Typical rebreather schematic. Carbon dioxide-laden gas exhaled by the diver passes through the scrubber, removing CO_2 before inhalation. Illustration courtesy of Jeffrey Bozanic and Best Publishing Company.

The work comes from several independent methods used by the author: laboratory-derived test data at NEDU, two independent mathematical modeling efforts, both probabilistic, and a so-called multiphysics[c] simulation of random processes within scrubber canisters. The revelations apply to all rebreathers, civilian or military, semiclosed or fully closed.

This book describes physical and physiological (human) factors influencing the variability of canister durations. It considers the statistical nature of rebreather canister duration measurements, emphasizing what engineers call the *propagation of error*.

In a simulated scrubber canister model, the random nature of CO_2 absorption and heat generation is affected by the instantaneous flow of heat and molecules, revealing the statistical nature of scrubber canisters. The simulated physical model lets us *see* what remains hidden within rebreather scrubber canisters' dark and caustic confines.

Carbon Dioxide Absorption

Any closed-circuit breathing apparatus depends on a CO_2 absorbing canister to remove CO_2 from the diver's exhaled breath before that breath is recirculated. The typical CO_2 absorbent chemical mixture in diving is sodalime (Sofnolime or Dräger DiveSorb Pro). The components of sodalime are calcium hydroxide ($Ca(OH)_2$, >75%), water (~17% - 20%), and sodium hydroxide (NaOH, 3%). The MSDS for DiveSorb Pro shows moisture varying from 12-20%.)

Sodalime granules are highly alkaline and absorb about 19% of their weight in CO_2. One hundred grams of sodalime can absorb

---

[c] *"Multiphysics is defined as the coupled processes or systems involving more than one simultaneously occurring physical fields and the studies of and knowledge about these processes and systems."* Wikipedia

approximately 15 L of CO_2 at atmospheric pressure[9] before the exit gas exceeds 1% CO_2.

The useful life of a CO_2 absorbent canister, or CO_2 scrubber, is estimated based on the statistical treatment of data[10] obtained during unmanned testing of the absorbent canister. The U.S. and foreign governments consider a CO_2 absorbent canister expended when the partial pressure of CO_2 in the canister effluent, the gas leaving the canister and inspired by the diver, rises to 0.507 kPa (0.5% Surface Equivalent Value, SEV) from nearly 0 kPa in thoroughly scrubbed gas[d]. It is also accepted that formal canister durations are determined by unmanned testing.

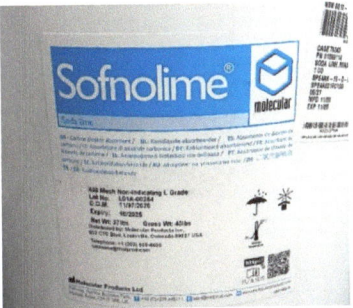

Figure 4. Pail of Molecular Products Sofnolime®.

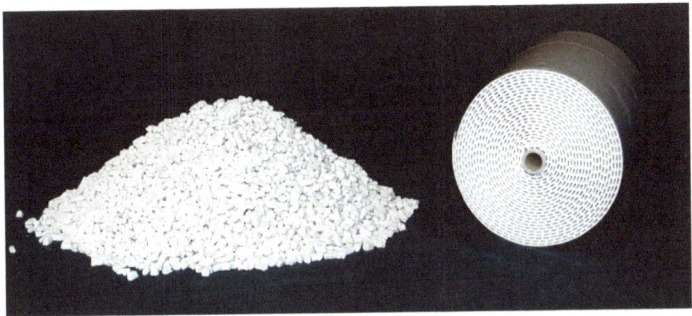

Figure 5. Sodalime CO_2 absorbent in granular and rolled forms.

---

[d] *The partial pressure of CO_2 in normal air is 0.04 kPa.*

According to the Micropore ExtendAir® Material Safety Data Sheet, those cartridges contain calcium hydroxide ($Ca(OH)_2$, >85%), sodium hydroxide (NaOH, <5%), and potassium hydroxide (KOH, <5%). (Commercial products frequently change, so consult the latest MSDS for updates.)

To provide the highest quality data with a minimal number of canister tests, unmanned testing employs fixed, constant test parameters: CO_2 injection rate into the canister and respiratory minute volume (RMV, in liters/minute or lpm[e]) from the breathing machine. In the real world, however, those parameters may be highly variable. This book addresses the physical and physiological variables that affect canister durations and discusses the implications for canister duration limits based on unmanned testing.

Figure 6. A rebreather scrubber canister filled with loose sodalime granules.

The primary question driving this analysis is as follows: if canister durations are determined from unmanned tests with their

---

[e] *In proper scientific usage, liters per minute is written as liter·min⁻¹ or L/min. As an aid to reading, throughout the rest of this book we will use the informal shorthand version, "lpm."*

controlled parameters and, therefore, minimal variability in measured canister durations, what impact does biological variability have on diver safety?

Figure 7. EX-19, the U.S. Navy's first digital rebreather. Photo credit: Bernie Campoli.

Or expressed another way, if canister limits are based on safe limits derived from unmanned tests, how risky are the highly variable man dives conducted to those limits?

Examination of the frequency distributions for multiple variables might answer these questions. It is not only the *mean* or average values for relevant variables in man dives that are important (e.g., oxygen consumption or CO_2 production) but also the scatter or variability of those variables. It's also important to know if those variables are distributed symmetrically about the mean, as in a *normal* or Gaussian distribution, or from a skewed distribution such as lognormal distributions.

Size Characteristics of Granular Sodalime

We sought to measure the granule size distributions for typical sodalime-based absorbents, samples of Sofnolime 812 and Sofnolime 408, small and large grain size absorbents made by Molecular Products, Thaxted, U.K.

A sample containing over 800 granules of each absorbent was photocopied, then scanned digitally. Automated image analysis extracted the number of granules and the size of each granule from the resulting image. That information was imported into a graphing program and then into a curve fitting software to find the "best" fit to the resulting size distributions of the granules (Figures 8 and 9.)

Details of that analytical process can be found in reference (11), available on line.

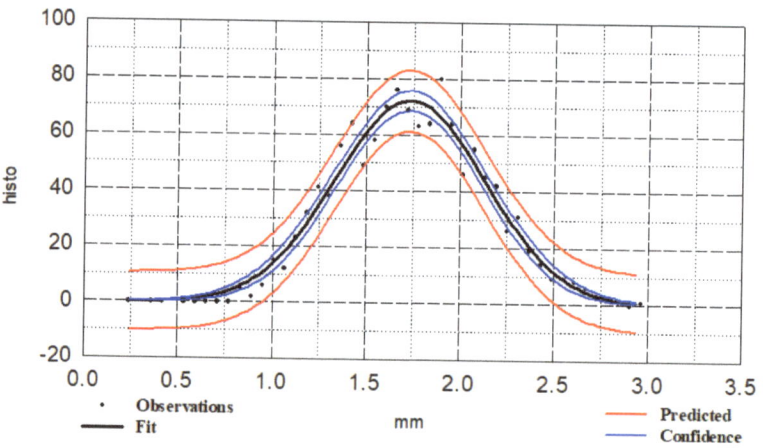

Figure 8. Confidence and prediction intervals on the Gaussian (normal) fit to Sofnolime 812[11].

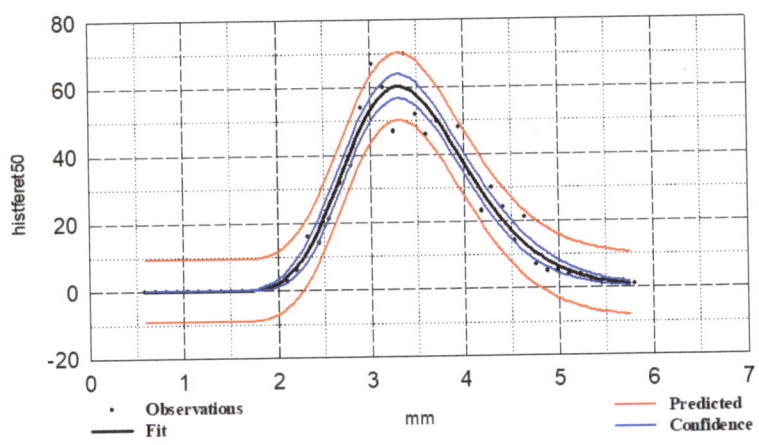

Figure 9. Log-Normal Fit to Sofnolime 408 granule size distribution.

As we'll see in the following chapters, information on these statistical distributions is far more critical than most divers realize. You will see how distributions are used to estimate the probabilities that a diver will encounter dangerously high levels of inspired CO_2 due to the exhaustion of a CO_2 scrubber.

Due to the variability in each of the multiple parameters involved in canister durations on manned dives, the resulting variability in canister duration is larger than estimates obtained in the laboratory. This is an unavoidable effect of the law of propagation of errors[f].

---

[f] *Notes on the Use of Propagation of Error Formulas, H.H. Ku, Journal of Research of the National Bureau of Standards, Vol. 70C, No. 4, 1966.*

Chapter 2. Background

Published canister durations rely on fixed laboratory test procedures, with the only variables being water temperature and canister packing. However, the useful lifetimes of scrubber canisters depend on many variables, some physiological and some physical. For instance, as a diver's CO_2 production rate ($\dot{V}_{CO_2}$)[g] increases, CO_2 begins exiting ("breaks through") the scrubber at shorter times, as seen in Figure 10). Four combat trained swimmers performed identical swims with rebreathers, but their times to breakthrough to 0.5 kPa and 1.0 kPa CO_2 varied with their oxygen metabolism.

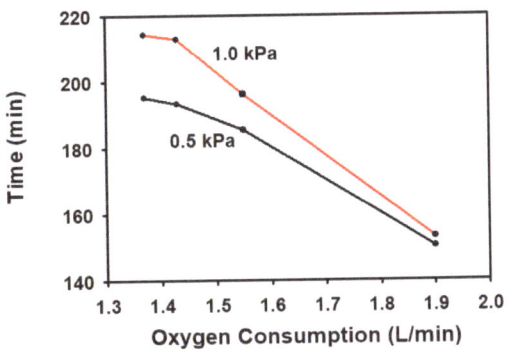

Figure 10. Identical swims with four fit divers produce varying canister breakthrough times.

---

[g] *The dot symbol over a variable is Newton's notation for a derivative. The first derivative of volume over time ($\dot{V}$) is flow, or change in volume over change in time. When V is expressed in liters, $\dot{V}$ has units of lpm. $\dot{V}_E$ is Respiratory Minute Volume (RMV) in lpm, and $\dot{V}_{CO_2}$ is CO_2 production in lpm at scientific standard temperature and pressure, dry (STPD).*

The divers' CO_2 production rates ranged from 1.37 to 1.9 lpm STPD, yielding a range of canister durations.

Breakthrough is defined as the elapsed time until the CO_2 partial pressure in the canister effluent reaches 0.51 kPa (3.8 mmHg or 0.5% surface equivalent value, SEV).

In one series of American dives by Knafelc[12], NEDU divers' measured oxygen consumption in combat swimmers averaged 1.52 ± 0.20 lpm (mean ± standard deviation, S.D.), range (1.21 - 1.77 lpm).

As this book shows, oxygen consumption and CO_2 production are not the only sources of variability influencing the risk that a diver will be exposed to high CO_2 levels while following supposedly safe published canister durations.

Unmanned Canister Duration Tests

Canister durations are determined in an unmanned laboratory where the major variables that might affect canister duration on an actual dive are controlled. Canisters are packed by humans, but everything else is handled by machines. According to NEDU Technical Manual 15-01[6], metabolic CO_2 production ($\dot{V}_{CO_2}$) is fixed at 1.35 liter·min^{-1} based on an oxygen consumption ($\dot{V}_{O_2}$) of 1.5 liters·min^{-1} and a respiratory exchange ratio (R or RER) of 0.9. The respiratory exchange ratio is the ratio of $\dot{V}_{CO_2}$ to $\dot{V}_{O_2}$.

RMV or $\dot{V}_E$ is fixed at 40 lpm throughout the laboratory test. Consequently, residence time (t_r), the time that exhaled CO_2 remains in contact with CO_2-absorbing granules within the scrubber canister, is constant.

Figure 11. NEDU's Unmanned Testing Laboratory and chambers. Copyrighted photos courtesy of Stephen Frink.

A Military Oxygen Rebreather

Figure 12. A Warrant Officer Navy SEAL with a U.S. Navy Oxygen rebreather.

An oxygen rebreather is the simplest fully closed rebreather. There are no electronics controlling oxygen addition. As the diver consumes the oxygen in the breathing bag (the brown bag in Figure 13), the bag gets smaller. Oxygen is only added when the breathing bag collapses and activates an oxygen addition valve.

Figure 13. The backside of a LAR V (MK 25). The oxygen bottle, regulator, breathing bag, hoses, and mouthpiece are visible.

Like all rebreathers, the LAR V has a CO_2 scrubber canister. Like all rebreather scrubbers, its function during a dive is a bit of a mystery. The following information in this monograph will help dispel some of that mystery.

Residence Time

The concept of CO_2 residence time within a scrubber canister is simple. The more time a CO_2 molecule spends in the scrubber, the more likely it is to be absorbed. According to a W.R. Grace publication[9], when residence time is less than 1 second, CO_2 absorption capacity is greatly reduced.

The concept of residence time may be simple, but some of the factors that influence residence time are anything but simple.

The prime mover in a rebreather is a diver's lungs. The harder the diver breathes, the faster gas travels through the UBA, including the canister. And therefore, less time is spent in the scrubber canister.

Equally noticeable is the effect of canister size. For a given average flow rate, the smaller the canister, the shorter the residence time within the scrubber when considering the bulk flow of gas through the canister.

The arrangement of breathing bags has a considerable influence. Gas flows through the scrubber of a one-bag rebreather only during exhalation and then at a high rate, reducing CO_2 residence time within the scrubber. When the instantaneous flow rate is calculated for a two-breathing bag system versus a one-breathing bag rebreather, the two-bag system markedly lowered peak flow rates[13]. That means that instantaneous CO_2 residence time within the scrubber is more consistent in the two-bag design (Figure 14) than in the one-bag system.

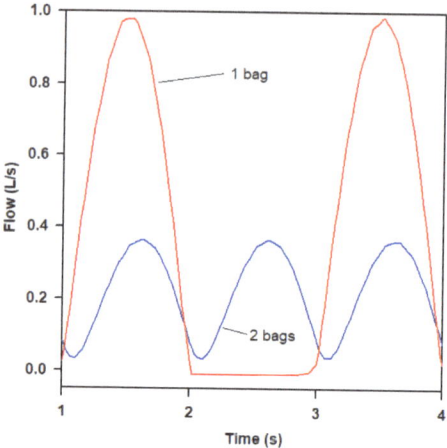

Figure 14. The calculated effect on gas flow waveforms through a canister in a two-bag rebreather versus a single-bag rebreather.

Less apparent but more critical is the porosity of the packed absorbent bed. The packing efficiency of the bed has an influence, but the effect of sodalime granule size distributions. The greater the spread of granule sizes in any given sodalime, the smaller the available flow paths through the canister. Smaller flow paths increase breathing resistance and cause higher velocity flow through the absorbent bed. That means residence time is reduced compared to narrow granule size distributions.

As described by Lin and Jaeger[14], the residence time is found by multiplying scrubber volume by packed bed porosity and dividing by volumetric gas flow rate. Further details of the influence of absorbent bed porosity are seen in Chapters 5 and 6.

Chapter 3. Prior Research

Described below is some of the early work of relevance to the theme of this monograph. For obvious reasons, I have not attempted to cover the entire body of knowledge.

U.S Navy, Washington D.C.

It seems to be a universal fact that divers are creative and self-confident. That combination of traits can lead divers to innovate "improvements" upon the work of life-support engineers. Sometimes those changes are beneficial, and sometimes they can be insidiously dangerous.

A case in point is the 1960s concept of reducing the absorbent load in a rebreather canister to match the anticipated dive duration. For instance, take an oxygen rebreather canister designed to hold five pounds of Baralyme[h] absorbent for a dive duration of two hours. If you only intend a 1-hour dive, reduce the absorbent load by 50% and fill the space in the canister with inert material, such as polystyrene rods.

A physician, LCDR M.W. Goodman, from the Navy Experimental Diving Unit was skeptical of the "reduced canister load" concept and tested it in the best scientific manner he could

---

[h] *Baralyme was a mixture of 80% calcium hydroxide and 20% barium hydroxide. It was used in anesthesia machines, early rebreathers, and during the SeaLab project.*

design. His tests were unmanned in a laboratory and manned with divers in open water swims.

Goodman's observations in a small group of divers caused him great concern. With equally fit Navy divers, indistinguishable in every physical and experience respect that a physician could conceive of in the 1960s, and tested under the same speed, distance, and environmental conditions, some of his divers had a canister duration 50% that of other divers. That result was wholly unanticipated and troubling.

Section 4.3 of Goodman's NEDU Report[15] was titled <u>Unanticipated Aspects of the Experimental Data</u>. His comments were sage, well-written, and foreshadowed one of the overarching themes of this current monograph.

> "The range of variation, 'individual variability,' occurring in the underwater swim tests far exceeded the anticipated magnitude. Presumably, we are hereafter obligated to assume that, should the test population be sufficiently large, some examples of absolute canister exhaustion will occur very early in the time span of the stress. It is axiomatic that individuals are, after all, individual... It is recognized that some healthy people consistently respond to stresses in a manner or degree which is abnormal, i.e., outside the "normal range."...It seems inescapable to conclude that sooner or later the chance meeting of a reduced-load canister which fails too rapidly, with a diver who exhibits responses outside the "normal range" will occur. A fatality may result. This conclusion, of course, leads to an equally inescapable recommendation: there is no justification whatsoever for utilization of the reduced canister load concept."

One of Goodman's recommendations was "that a project be established to authorize basic studies of absorption system design, with the expectation that the dimensions and specifications for an optimal system will result therefrom."

To that end, one of the most detailed engineering writings on CO_2 scrubber canisters were by Nuckols, Purer, and Deason, published in original and revised version in 1983 and 1985. Based in Panama City, Florida, they created *Design Guidelines for Carbon Dioxide Scrubbers*[16]. Naval Coastal Systems Center (NCSC) Technical Manual 4110-1-83 can be downloaded from the internet and is well worth perusing[i].

Institute of Naval Medicine, Yugoslavia

In 1987, Radić and colleagues measured the effect of diver ventilatory patterns and CO_2 injection rate on scrubber duration[17] using a 2 kg load of DrägerSorb 800 absorbent in a 2.2 L scrubber canister attached to a prototype rebreather. The test apparatus was submerged in a 20°C water bath under atmospheric pressure and ventilated at two flow rates, 27 and 53 lpm.

The raw data from Radić et al. were replotted in Figure 15 of this document. Using nonlinear regression, this author determined that the fitted curves (blue for 27 lpm and red for 53 lpm) match the curves of an inverse, second-order polynomial,

$$f = yo + \frac{a}{x} + \frac{b}{x^2}$$

where f is duration, x is CO_2 injection rate, and a and b are constants.

---

[i] *Appendix D in this monograph introduces an alternative version of an equation in reference (16), a version that is relevant to the findings of this current work.*

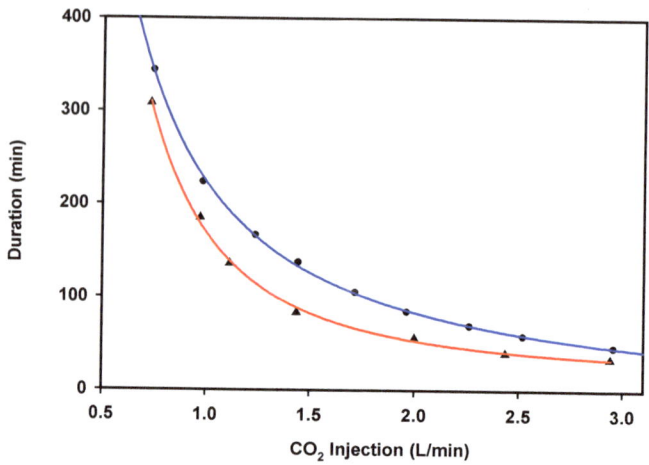

Figure 15. Studies of the effect of ventilation parameters on CO_2 scrubbers[17]. Blue curve = $\dot{V}_E$ of 27 lpm, red curve = $\dot{V}_E$ of 53 lpm.

It is not surprising that the higher the CO_2 injection, the shorter the canister duration. What is not always appreciated is that the harder a diver breathes for a given CO_2 production ($\dot{V}_{CO_2}$), the shorter the canister duration.

Little attention has been paid to the marked curvilinearity of these $\dot{V}_{CO_2}$ curves. In fact, those curves are rarely found in the literature. However, the curvilinearity of these curves will be featured later in this monograph.

Radić et al. also found that for an average $\dot{V}_{CO_2}$ of about 1.5 lpm (1.47 ± 0.005 lpm, n=16), the amount of CO_2 absorbed before breakthrough depended not only on $\dot{V}_E$ but also on tidal volume. For a one-liter tidal volume, CO_2 absorbed ranged from 73 L/kg to 102 L/kg. For a two-liter tidal volume, CO_2 absorbed varied from 60 L/kg to 101 L/kg.

University of Florida

A decade after the publication of the U.S. Navy design manual, Marc Jaeger, an M.D. respiratory physiologist and an engineer colleague M-J Lin, produced one of the most interesting studies of sodalime canister performance relevant to diving. In 1994, their paper *CO₂ binding by Baralyme in three different carrier gases*[14] was published in Undersea and Hyperbaric Medicine.

Lin and Jaeger's theory of scrubber operation can be reduced to the following: T defines the time for an absorption reaction to end because all the absorbent has been consumed. T in minutes is defined as,

$$T(min) = \frac{absorbent\ quantity\ (moles)}{inflowing\ CO_2\ (\frac{moles}{min})}$$

This is a common assumption discussed further in Appendix D. It certainly makes intuitive sense. However, as we will see throughout this book, *the devil lies in the details.* Those details will be illustrated in Appendix E.

The primary findings of Lin and Jaeger's studies at 1 atmosphere absolute (1 ATA) were as follows:

1) The life of a scrubber is longer when the inlet CO_2 concentration is high, and the gas flow rate is low.
2) Scrubber life is longer in *helium* than in heavier gases like *oxygen* and *sulfur hexafluoride*.
3) Contact (resident) time is critical.
4) The amount of bound CO_2 in percent capacity is closely related to the ratio of *breakthrough time* (to 0.5% SEV) and *residence time*.

NEDU

NEDU's contribution during 1979 was to conduct the deepest ever Navy heliox saturation dive to 1800 fsw. Logistics for the dive was daunting. It was a 38-day dive with six divers. If we assume that each diver exhaled an average between work and rest of 1 lpm of CO_2, that would yield 328,300 L or 645 kg of CO_2 during the dive. Assuming a typical 50% absorption efficiency for sodalime, 1.4 tons of absorbent would be needed. That translates to 77 37-lb pails of Sodasorb or 65 20-kg pails of 797 Sofnolime.

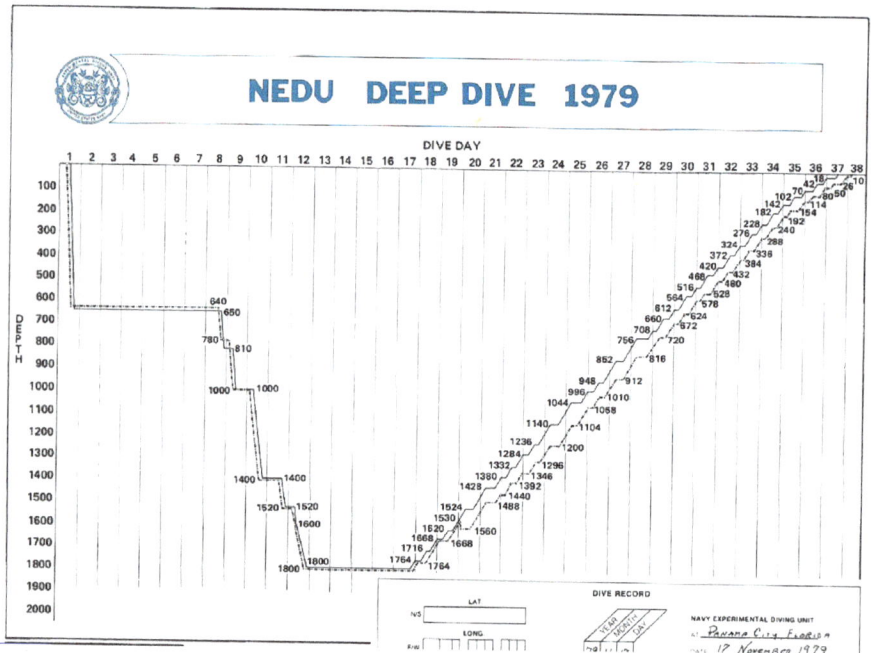

Figure 16. The U.S. Navy's deepest heliox dive at NEDU. From Faceplate, Naval Sea System Command, Washington.

Chapter 4. Mathematical Models

Are you a child of Lake Wobegon?

In the now-defunct radio show, *A Prairie Home Companion,* public radio star and author Garrison Keillor delivered the news from a fictional small town in central Minnesota named Lake Wobegon. Lake Wobegon's claim to fame was that "all the women are strong, all the men are good-looking, and all the children are above average."

If you were a child of Lake Wobegon, what would be the consequences of being "above average?" Of course, that depends on the question being asked. The central question regarding rebreather scrubber canisters is, when will the canister break through? When will I have to abort my dive?

The answer to that question depends on whether, on any given day, on any given dive, your dive gear and your bodily functions are average, below average, or above average. The truth is, it's not always good to be "above average."

This chapter explores in detail that simple fact of statistics.

Approach

A Dräger manual for a semiclosed rebreather like the Dolphins on this book's front cover asserts that it has a "180-minute" canister. That sounds comforting if you plan a 120-minute dive. But can you plan on getting 180 minutes out of that canister? Well, the salient point of this entire chapter is, don't bet your life on it!

As any rebreather testing laboratory will tell you, inhaled CO_2 remains near zero for most of the canister run if there is no channeling or short-circuiting of the absorbent bed. Eventually, inspired CO_2 rises off the zero baseline and begins a steadily increasing climb upwards. That upward curve invites comparison to an exponential curve.

While the overall shape of a canister breakthrough curve appears simple at first glance, it hides a multitude of factors that influence the exact shape and timing of that curve. Regrettably, to appreciate the influence of those factors in toto, we must delve deeper into math.

The reward for this effort is that the determined reader will understand the effect of human and equipment variability on the potential for an untoward dive outcome—i.e., a bad dive.

Specifics

As a reminder, canister *breakthrough* refers to CO_2 molecules escaping capture by the absorbent granules and recirculating through the diver's breathing circuit. Three hours or 180 minutes is the quoted canister time allotted for one of the most ubiquitous electronic closed-circuit rebreathers, the Inspiration (XPD and EVP), using a 2.5 kg scrubber absorbent load. Three hours is also the stated canister time for the Dräger Dolphin semiclosed rebreather and, conveniently, is the NOAA oxygen exposure limit for a 1.3-atmosphere PO_2 dive.

Figure 17 shows CO_2 breaking through a generic CO_2 canister within 180 minutes. For reasons explained below, the CO_2 concentration rises in the canister effluent and, therefore, in the diver's inspired gas, at an accelerating pace following the breakthrough. The escaping CO_2 keeps reentering the loop, a prescription for exponential growth.

The modeled exponential fit (red line) to the data from four canisters was determined using TableCurve 2D by Systat Software, Inc., Richmond, CA.

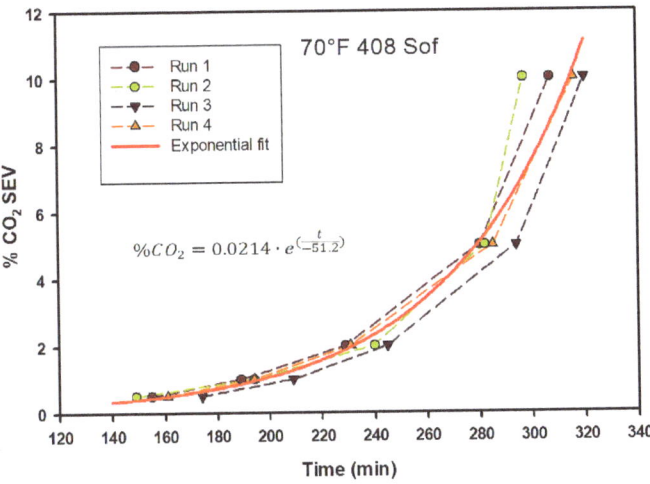

Figure 17. Four "identical" canister duration runs in 70°F water.

According to the U.S. Navy protocol, CO_2 was injected at 1.35 lpm and a ventilation rate of 40 lpm. The Navy protocol also requires five canister runs. Figure 18 shows the fifth run of that test series.

The fifth scrubber (blue curve) broke through early when an unspecified amount of water leaked into the canister. However, complete flooding did not occur since the water was not sucked into the gas analysis system.

The canister breakthrough curve has two components. The first component typically lies along the zero CO_2 axis (not shown), where all CO_2 entering the canister is absorbed, leaving no CO_2 in the effluent. After some time t_{th}, called the threshold time, CO_2 rose steeply.

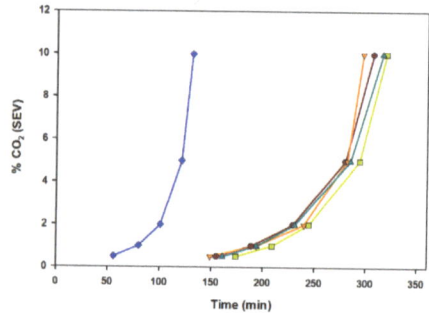

Figure 18. A fifth canister (blue curve) broke through early due to water intrusion.

Derivation of Breakthrough Equation

If you're a rebreather diver reading this book, you understand the basics of human physiology. Our bodies breathe in oxygen and produce carbon dioxide as a waste product. One purpose of the lungs is to remove CO_2 from the body and deliver it to the atmosphere. Likewise, the purpose of the rebreather scrubber is to remove CO_2 from the rebreather, an extension of our lungs. Unlike our atmosphere, a scrubber's CO_2 absorption capacity is limited.

A breakthrough equation puts numbers to that last statement.

Alveolar Gas Equations

Throughout our diving careers, we have all experienced shallow and slow breathing when photographing an animal underwater and increased breathing when swimming. The alveolar gas equation directly relates the level of ventilation to respiratory CO_2 production.

The ventilation of the gas exchanging, alveolar regions of the lungs, $\dot{V}_A$, can be described by,

$$\dot{V}_A = 863 \, mmHg \cdot \frac{\dot{V}_{CO_2}}{P_{aCO_2}}$$

where $\dot{V}_A$ is alveolar ventilation in lpm, $\dot{V}_{CO_2}$ is CO2 production rate in lpm, and P_{aCO_2} is the partial pressure of CO2 in arterial blood.

If $\dot{V}_{CO_2}$ is 1.4 lpm, and arterial blood P_{aCO_2} is its expected value of 40 mmHg, then $\dot{V}_A$ would be 30.2 lpm.

An alternative version of this equation is,

$$\dot{V}_A = Rgas \cdot \dot{M} \cdot \frac{T}{P_{aCO_2}},$$

where Rgas[j] is the gas constant, mass flow in moles per minute is $\dot{M} = \frac{\dot{V}_{CO_2}}{Vm}$, and Vm is the molar volume for perfect gases, 22.414 L/mol at STP conditions (1-atmosphere pressure, dry and 0°C.) T is the alveolar temperature in Celsius, typically assumed to be 37°C.

Total ventilation is the combination of alveolar and dead space ventilation, $\dot{V}_E = \dot{V}_A + \dot{V}_D$. $\dot{V}_D$ is both physiological in origin, occurring within human airways, and mechanical in origin, found in full face masks, helmets, and rebreather mouthpieces. $\dot{V}_D$ exists anywhere CO2 can accumulate and can easily contribute 8 lpm of added ventilation to our example above.

We defined the canister breakthrough threshold as a direct function of the mass of sodalime in the scrubber canister and the residence time of gaseous CO2 flowing through the sodalime absorbent bed. It is inversely proportional to the rate of CO2 production, $\dot{V}_{CO_2}$. $\dot{V}_{CO_2}$ is calculated by multiplying oxygen consumption, $\dot{V}_{O_2}$ in lpm STPD, by the respiratory exchange ratio (R).[k] For our calculations, we assumed R = 0.9.

$$\dot{V}_{CO_2} = \dot{V}_{O_2} \cdot R \tag{1}$$

---

[j] *The universal gas constant is $8.31446 \, J \cdot mol^{-1} \cdot K^{-1}$*

[k] *R is a more concise symbol for RER, and for that reason is preferred in equations.*

For each pass through a canister, the time a CO_2 molecule remained within a canister, or CO_2 residence time (t_r), was inversely proportional to minute ventilation $\dot{V}_E$.

$$t_r = \frac{V_s \cdot \epsilon}{\dot{V}_E} \qquad (2)$$

In equation (2) from Lin and Jaeger (1994), Vs is the scrubber canister volume in liters, and ϵ is the porosity of the absorbent bed. For a given minute ventilation, the lower the bed porosity, the faster the CO_2 molecules must pass through the canister. As a result, the CO_2 residence time is decreased.

A different but equivalent equation for residence time is found in Nuckols, Purer, and Deason (1983, with correction in 1985,)

$$t_r = \frac{Lcan}{V}$$

where *Lcan* is the length of the scrubber canister, and *V* is superficial gas stream velocity.

$$V = \frac{\dot{V}_E}{Acs \cdot \epsilon}$$

Acs is the cross-sectional area of the canister, and, again, ϵ is absorbent bed porosity.

The proportionality between ventilation ($\dot{V}_E$) and Pa_{CO2} is controlled by the diver's ventilatory sensitivity to arterial CO_2. K_{CO2} is the ratio between ventilation and $\dot{V}_{CO_2}$.

$$K_{CO_2} = \frac{\dot{V}_E}{\dot{V}_{CO_2}} \qquad (3)$$

For this notional model, the probability (P) of an absorption reaction was assumed to be affected by temperature in the following manner:

$$P = \frac{T - 28°F}{T - 14°F}$$

For an absorbent temperature of 70°F, P is 0.75, a value that enters into the estimation of threshold time[1].

The CO_2 absorbency of sodalime is known to vary between 10 and 26 L of CO_2 STPD for every 100 grams of Sodasorb. We chose the most conservative value of 10 L/100 grams to begin our estimation of the theoretical absorbency (abs) for the canister.

$$abs = \frac{10L \cdot mass}{100 \cdot gm}$$

For a 2.2 kg (5-pound) mass of absorbent, the theoretical minimum volume of CO_2 absorbed at STP conditions was 227 L.

The number of CO_2 absorption sites available per second is not knowable. However, the ratio of absorption sites can be inferred from granule size and the number of granules. That exercise is completed in Chapter 5.

Borrowing from those calculations, we assert the following: The ratio of total absorption sites in Sofnolime 812 (fine grain) absorbent is approximately 1.33 times the number of sites in large grain size absorbent (Sofnolime 408.) Accordingly, when running an estimate for fine grain absorbent, we assume a fine grain/large grain ratio (S) of 1.33/sec. When doing the same for large grain sodalime, S = 1.0/sec. Therefore, S becomes another estimate of absorption efficiency.

The canister threshold time (t_{th}) can be modeled by,

$$t_{th} = \frac{P \cdot S \cdot abs \cdot t_r}{\dot{V}_{CO_2}}. \qquad (4)$$

---

[1] 28°F is the freezing temperature of seawater found under Arctic and Antarctic ice. 14°F is a factor in this hyperbolic equation increasing P with warming.

Equation (4) includes the effect of scrubber bed temperature, the use of fine or large grain sodalime, the total mass of absorbent, CO_2 residence time within the scrubber, and of course, the CO_2 production rate.

The notional model for CO_2 breakthrough is now found by combining Equations (1 – 4) into Equation (5), representing the partial pressure of CO_2 in the diver's inspired gas.

$$P_{ICO_2} = a1 \cdot e^{\frac{a2 \cdot \dot{V}_{CO_2}(t-t_{th})}{a4}} \qquad (5)$$

The I in P_I stands for *inspired*. Once the threshold time (t_{th}) was exceeded, the partial pressure of inspired CO_2 in the canister effluent rose exponentially, with a rate of rise proportional to $\dot{V}_{CO_2}$.

The results from applying equation (5) are found in Table 1. The qualitative effects of increasing temperature are located in rows 1 and 2. The result of increasing oxygen consumption and CO_2 production is found in row 3. Using large grain absorbent instead of fine grain absorbent (row 4) reduced the previously estimated threshold by 25% or 39 minutes.

Increasing the amount of absorbent and diving with a resting workload and ventilation are shown in rows 5 and 6, respectively. *(Arrows indicate the direction of change moving from top to bottom within the table.)*

Table 1. Scrubber canister simulation.

Row	T °C	$\dot{V}_{O_2}$ lpm	$\dot{V}_{CO_2}$ lpm	$\dot{V}_E$ lpm	K_{O_2}	K_{CO_2}	Mass kg	S	tr sec	tth min
1	-1.1	1.5	1.35	40	26.7	29.6	2.3	1.33	1.13	32
2	21.1	1.5	1.35	40	26.7	29.6	2.3	1.33	1.13	189↑
3	21.1	1.8↑	1.62	40	22.2	24.7	2.3	1.33	1.13	158↓
4	21.1	1.8	1.62	40	26.7	29.6	2.3	1.00↓	1.13	119↓
5	21.1	1.5	1.35	40	26.7	29.6	3.6↑	1.33	1.13	254↑
6	21.1	1.0↓	0.9↓	22↓	22.0	24.4	2.3	1.33	2.06↑	517↑

K_{O2} was the ventilatory equivalent for oxygen, and K_{CO2} was the ventilatory equivalent for CO_2. R = 0.9 and absorbent bed porosity = 0.32 for all rows.

Equation 5 produces CO_2 breakthrough curves like Figure 19, which plots the results from row 2 of Table 1. In this instance, the threshold time tth was 189 min. Breakthrough (BT) to 0.05 kPa (or % SEV) occurred at 202 min.

Equation (5) is not meant to be a quantitative predictor of canister duration for any rebreather. It only points out the number and type of variables involved in producing a CO_2 breakthrough curve in Figure 19.

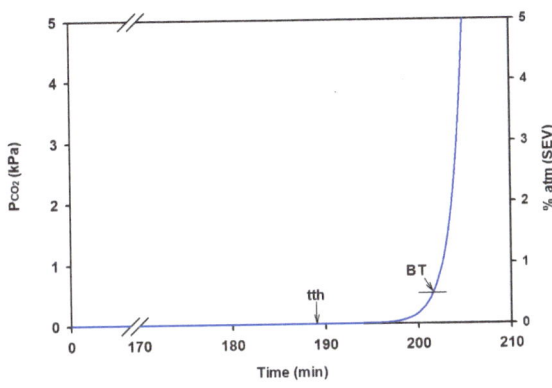

Figure 19. CO_2 breakthrough curve from equation (5).

Variables

Having defined the variables in the breakthrough equation, we now explore the influence of uncertainty in each variable. That is the focus of the remainder of this chapter.

The variables within the mathematical model were distributed as follows:

1) The mass of sodalime packed into the canister. While this is a physical quantity, it is biological in origin because the canister is packed by a human who invariably uses slightly different packing procedures for each absorbent fill. Data for sodalime packing comes from a Canadian study that measured the weight of sodalime in rebreather canisters as packed by highly motivated operators. This error source becomes less critical in pre-formed absorbents such as Micropore's ExtendAir cartridges.
2) Oxygen consumption. Even when the workload of a freely swimming diver is tightly constrained, as in a combat swim at a sustainable pace in a no-current condition, there is invariably a spread of oxygen consumption. The difference between the greatest and least oxygen consumption can be considerable due to differing body sizes, swimming speed, and swimming efficiency. We primarily used data for this analysis from combat swimmers[3].
3) Respiratory exchange ratio, R. R varies with the intensity of work and the substrate used for metabolic oxidation (fat or carbohydrates or a mixed diet.)
4) Ventilatory sensitivity to CO_2. As workload and $\dot{V}_{CO_2}$ increase, ventilation must increase to maintain arterial CO_2 within normal limits. This increase in ventilation reduces the residence time (t_r) of exhaled CO_2 molecules within the scrubber canister. However, some divers ventilate less than other divers in response to the same $\dot{V}_{O_2}$, resulting in a higher arterial CO_2 level but lower respiratory pressures and a feeling of easier breathing. As a side effect, slower breathing allows longer CO_2 residence times within the canister and, therefore, more efficient canister scrubbing. This ventilatory sensitivity is expressed as K_{CO_2} in Equation (3) and is

allowed to vary, i.e., be distributed over an experimentally measured range of values.

Unmanned Testing Variation

Before measuring canister durations in military rebreathers, the test absorbent passed through prequalification testing[5,18,] which verified that samples from a given lot of absorbent met the manufacturer's specifications for granule size distribution, moisture content, and CO_2 absorbency. (As explained in Chapter 1, manufacturing variability has been a major issue and is likely to be an additional source of variability in the future.)

Even with careful controls on testing conditions, procedures, and absorbent quality, a considerable canister duration variance existed in tested rebreathers, as shown in Table 2. The coefficients of variation, or COV, is the standard deviation of duration divided by mean time to reach various CO_2 effluent concentrations. Table 2 shows COVs from low to high CO_2 measured at various points along the CO_2 breakthrough curve (0.5 – 2.0% CO_2).

Table 2. Unmanned coefficients of variation (COV).

Absorbent	n	fsw	°F	COV 0.5% CO_2	COV 1.0% CO_2	COV 1.5% CO_2	COV 2.0% CO_2	COV Weight%
Sof 812	7	190	29	16.3%	11.4%	8.0%	6.4%	1.7%
Sof 812	5	190	50	2.5	2.9	3.0	3.1	--
Sof 812	5	190	70	3.9	2.8	2.0	1.7	2.1
Sof 812	5	190	90	4.0	3.0	2.2	1.8	1.2
Sof 812	2	20	105	3.0	2.7	2.9	3.1	--
Sof 408	5	190	29	11.2	13.0	12.9	12.2	2.2
Sof 408	5	190	50	3.1	2.6	3.4	3.8	1.7
Sof 408	5	300	50	9.0	8.0	8.2	8.3	1.8

Sofnolime 812, Sof 408 = Sofnolime 408

The COV for canister duration testing tended to be the greatest under the most extreme conditions, e.g., 29°F water temperature (16.3% and 11.2%) and deepest depth (300 fsw, 9.0%).

Physiological Factor Variation

Canister duration during a manned dive depends at least as much on variation in physiological factors as it does on the physical factors affecting unmanned testing. In unmanned testing, conditions are constrained as much as possible by depth, CO_2 injection rate, water temperature, and ventilation rate. But still, variation occurs.

In manned testing, the variability of human physiology cannot be tightly controlled, even if the test conditions are constrained by age, gender, physical and medical fitness, and diving experience.

A confounding factor in human dives is that sometimes two of the most important variables affecting canister duration, R and $\dot{V}_{O_2}$ interact. Sometimes they cannot be treated as independent factors due to a strong covariance, but sometimes they can.

For instance, two contemporaneous studies by well-respected research teams asked a simple question: To what degree is the respiratory exchange ratio (R) correlated with oxygen consumption ($\dot{V}_{O_2}$)? Well, the answer depends on who you ask. In the Dwyer and Pilmanis study[19], the association between R and oxygen consumption is pronounced, while in Morrison's work[20], it is relatively weak.

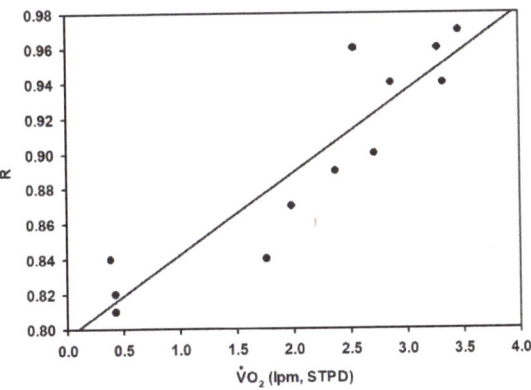

Figure 20. Dwyer and Pilmanis[19]. Only mean values were provided.

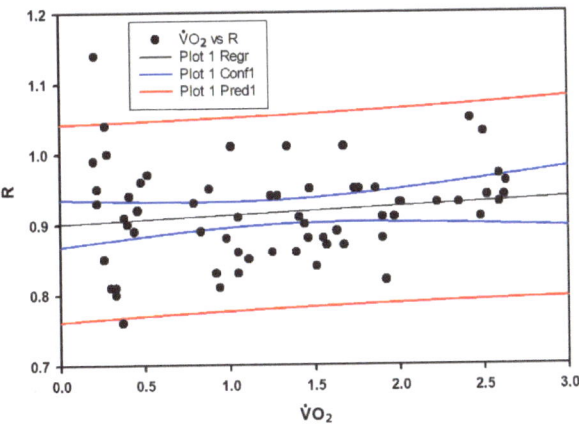

Figure 21. Morrison's correlation (covariance) between R and oxygen consumption[20].

A strength of the Morrison study is that it has enough data to calculate confidence limits on the best estimate of R as a function of oxygen consumption. It also allowed the calculation of the 95% prediction limits for R.

Figure 22 shows the implications of variation in the mass of sodalime absorbent packed in a military oxygen rebreather. Likewise, Figures 23-25 reveal potential distributions of oxygen consumption ($\dot{V}_{O_2}$), respiratory quotient (R), and the ventilatory equivalent for $\dot{V}_{O_2}$, namely $\dot{V}_E/\dot{V}_{O_2}$ or K_{O_2}.

Human Data Distributions

Each "bell curve" in Figures 22-25 shows a presumed normal distribution based on each parameter's measured mean and standard deviation. The source of the plotted data is listed below each graph.

Absorbent Mass

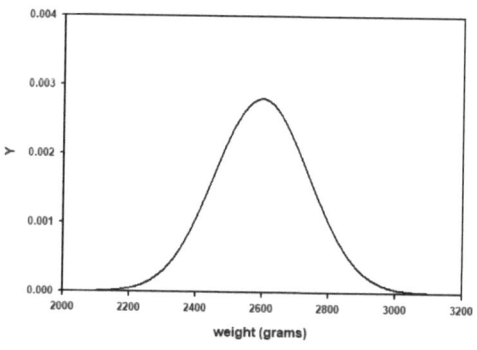

Figure 22. Mass of absorbent in an oxygen rebreather canister.

The average mass of absorbent in 41 packed canisters was 2597 ± 142 grams (mean ± SD).

One large and surprising source of *biological* variation was uncovered when well-trained Navy divers accustomed to packing LAR V (MK 25) canisters, packed the scrubber canister with granular absorbent (Sofnolime 812) for a new semiclosed UBA being evaluated at NEDU[21]. Using their conventional packing method, they produced a canister weighing 3.0 ± 0.1 kg (6.8 ± 0.2 lb.) The manufacturer had recommended in their Operations and Maintenance Manual a weight of 2.4 kg (5.5 lb.)

The breathing resistance of the *over-packed* canisters was tested by NEDU and found to be better than Navy performance specifications, and came close to meeting a contemporaneous resistive effort goal.

After that discovery, the manufacturer and program sponsor agreed to the new NEDU packing weight (24% higher than previously recommended) to help prevent channeling, and to prolong canister durations.

Translating that result to typical civilian rebreather operations, overfilling a canister will lead to decreases in absorbent bed porosity, and a resultant rise in flow resistance and resistive effort.

Assuming you <u>do not</u> have the means for measuring flow resistance in your canister under relevant conditions, and are unlikely to get manufacturer support for overfilling, the message should be clear. **Don't Overfill!**

Even if you do your own packing, and are diligent, canister packing is inherently subjective. Filled canister weights (and canister durations) will vary when using granular absorbent.

Oxygen Consumption

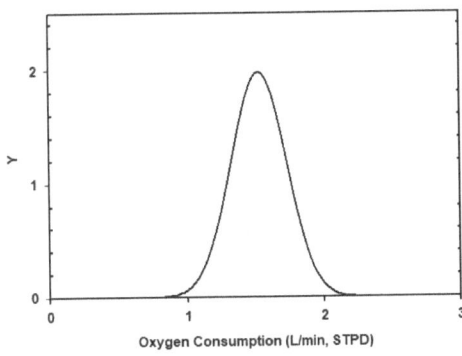

Figure 23. Variation of oxygen consumption, $\dot{V}_{O_2}$ (Knafelc[12]).

Knafelc measured a $\dot{V}_{O_2}$ of 1.52 ± 0.21 lpm (mean ± SD) in 8 divers performing combat swims, as shown in Figure 23.

Respiratory Exchange Ratio

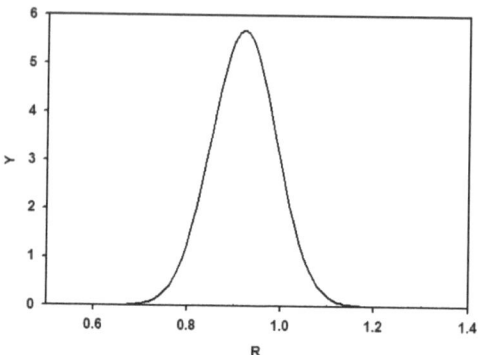

Figure 24. Variation of the respiratory exchange ratio, R.

Figure 24's data came from Morrison et al.[20]. R = 0.92 ± 0.07, n = 63.

Ventilatory Equivalent

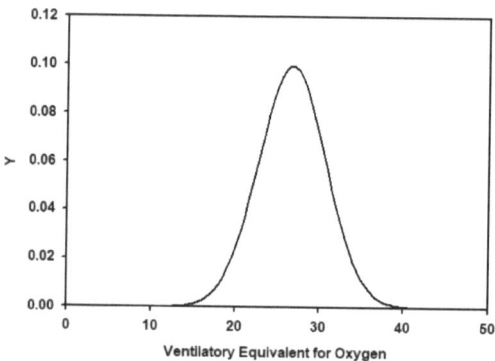

Figure 25. Variation of ventilation equivalent for oxygen, KO_2.

Figure 25 shows data from both NEDU and Morrison[20]. KO_2 = 26.7 ± 4, n = 63.

The COV for the experimental datasets for $\dot{V}_{O_2}$, R, KO_2 and absorbent mass were 13.8%, 7.6%, 15%, and 5.5%, respectively. These physiological COVs fall in the same range as the COVs for unmanned canister durations (Table 2).

Sampling Error

The data used to obtain information concerning the distribution of oxygen consumption, canister packing mass, etc., was based on relatively small sample sizes. Therefore, the sample's mean estimate did not accurately represent the population mean. There remains some uncertainty regarding the true mean value.

That is doubly true for the estimate of the population variance. Typically, the information obtained from a small sample size regarding a sample's central tendency is more accurate than information regarding the distribution's tails. We must remain uncertain about the actual probability of low-probability events.

Confidence intervals reflect our uncertainty in estimating a mean. That applies to either a univariate value, such as canister duration at a given water temperature, or a bivariate value, such as the regression of canister duration as a function of varying water temperatures. Unfortunately, in commercial publications for divers, confidence intervals are rarely used. That can lead to false confidence in published numbers, such as canister durations.

Prediction intervals (Figure 26) assess the variability of the total population behavior, not just the mean. Prediction intervals are always wider than confidence intervals on the mean.

In the U.S. Navy, when establishing canister durations as a function of water temperature, prediction intervals are used as the basis of the published limit[6]. However, before making that calculation, one must test whether the errors associated with each measurement are distributed normally in a Gaussian or bell-shaped fashion. Some statistical tests allow that.

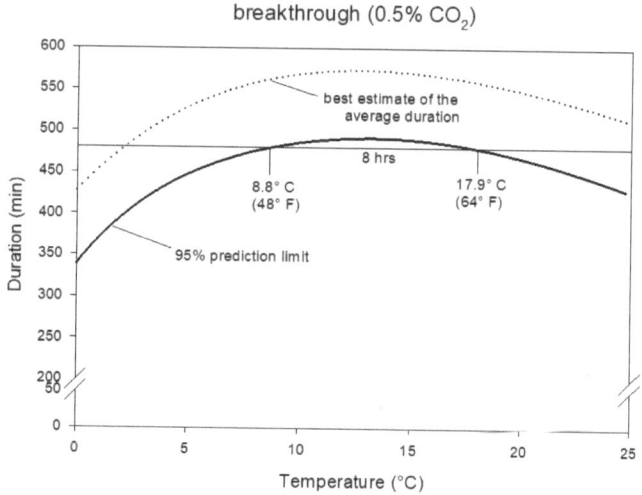

Figure 26. The relation between average canister durations and 95% prediction estimates. From Reference 10.

If the sampled data's errors seem skewed, the assumption of normally distributed errors may not be valid. In that case, prediction intervals calculated using that assumption may be erroneous. Unfortunately, it would be difficult to estimate how great that error might be.

When examining data on oxygen consumption, for example, the sample size is typically too low to give helpful information on either the normality of the data or the normality of the variance of the data. For some parameters, the normality assumption would be illogical, at least for the data.

For instance, based on first principles, skewed distributions such as log-normal distributions might be expected. Despite skewing of the data, the errors or variance associated with those distributions may well be normally distributed. Still, the number of samples would have to be large to prove that. We will not employ confidence and prediction intervals for the SPM-based canister durations of Chapter 5 for all the above reasons.

Because of our uncertainty in the actual variance of the data and the small sample sizes, we should expect that our estimates will underestimate the real consequences of biological variability. The meaning of that statement for rebreather divers like yourself will become more apparent as we proceed.

We examined the frequency distribution of sampled data used in this modeling effort and performed normality tests. In all cases, we could not reject the assumption of normally distributed data. However, where that assumption led to physically unreasonable conclusions, we assumed a skewed distribution[m].

In the manner of stepwise regression, we built variation into our models stepwise, adding variables one at a time. That allows the interested reader to test the influence of each variable on the model. Fortunately, the method we use is robust enough that an assumption can be modified or a mean can be altered without altering the basic, qualitative conclusion of this analysis

Propagation of Error

As a reminder, canister *duration* is the time required for P_{CO_2} in the effluent from a single CO_2 absorbent canister to rise to 0.5 kPa (0.5% SEV), i.e., to "break through." On the other hand, canister *limits* are the published times that a diver is allowed to remain on a canister at a given water temperature.

---

[m] *Two competing mathematical models of a data set may be difficult to select based on a single "goodness of fit" estimate. However, if one goes to infinity when extrapolated outside of the data range, that is probably not a good model to depend on. It's not physically realistic.*

Canister durations of underwater breathing apparatus are determined from repetitive laboratory tests over a range of water temperatures but at fixed ventilation and CO_2 injection rates. In the U.S. Navy, published canister limits based on measured canister durations and their variability (as discussed above), are defined as the lower 95% prediction limits for canister breakthrough under the conditions tested.[10]

Here, we explore the consequences of diving to the published canister limits **when the assumptions used in determining duration are not applicable**. The riskiness of such behavior is revealed by applying a classic engineering method, the propagation of error.

In preparation for this second mathematical analysis, we employ a simpler equation than Equation (5) and its components. We begin with Equation (6).

$$Y = a1 \cdot e^{a2 \cdot \left[(\dot{V}_{CO_2} \cdot t) - \frac{a3 \cdot mass}{\dot{V}_E}\right]} \quad (6)$$

To use the distribution data from Figures 22-25, we make substitutions for $\dot{V}_{CO_2}$ and $\dot{V}_E$. The result is Equation (7). The distributed variables to be examined are R, $\dot{V}_{O_2}$, KO_2, and mass of absorbent.

$$Y = a1 \cdot e^{a2 \cdot \left[(R \cdot \dot{V}_{O_2} \cdot t) - \frac{a3 \cdot mass}{KO_2 \cdot \dot{V}_{O_2}}\right]} \quad (7)$$

Equation (7) was used as the basis for the computation of error propagation.

We use that process to predict the effect of diving not with the most probable value of mass, R, $\dot{V}_{O_2}$, and KO_2, but by the less likely values plotted in those figures.

In other words, it's okay to feel lucky before a dive, but you would be wise to consider that your luck may change. In that event, what could happen to your scrubber duration?

To minimize the pain and suffering of the reader, I have removed from this document the set of partial differential equations required for the propagation of error calculations. Those and other mathematical calculations and ruminations will be made available in an e-supplement to this book.

Figures 27-29 result from those propagation of error (POE) formulas on expected canister durations, assuming *normal* (in the statistical sense) variances.

The following three figures examine the risk of inhaling 2% CO_2. Two percent was chosen because there is general agreement that 2% CO_2 is not an acceptable CO_2 level for a rebreather. Beyond 2%, CO_2 rises dramatically in an exponential fashion.

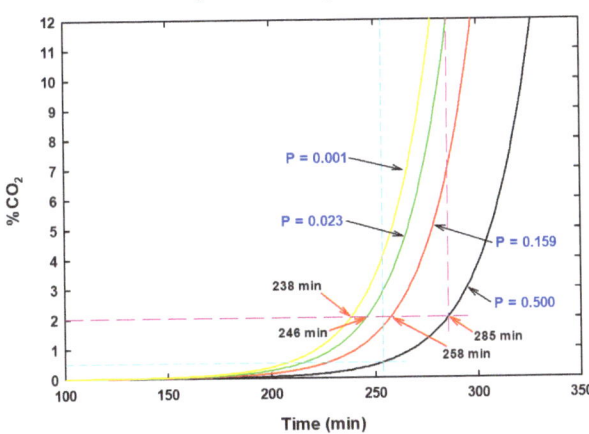

Figure 27. Breakthrough curves assuming a small covariance.

Figure 27 was based on data from Morrison et al.[20], with a small covariance between R and $\dot{V}_{O_2}$ = 0.007. The times (black lettering) and probabilities (blue lettering) of reaching 2% CO_2 are indicated for the four curves.

The black curve in Figure 27 is the average (P = 0.5) CO_2 breakthrough curve from the fitted "physiological" equation (7). The less probable curves are obtained from the propagation of error analysis.

As shown by the intersection of the dashed blue lines with the black curve in Figure 27, a simulated canister is expected to break through to 0.5% SEV at 253 minutes. Due to the propagation of error regarding the physiological uncertainty between a diver's oxygen consumption and his amount of CO_2 produced (R), equation 7 predicts that in 17 out of 100 dives, inspired CO_2 could be approaching 2%, 4 times the expected CO_2 load.

According to the math, there is a 2% chance that inspired CO_2 could reach 2% at 246 minutes into the dive. There is a one in 0.001 probability that it could be 2% at 238 minutes, 15 minutes before the expected expenditure of the canister.

That assumes that the covariance between R and $\dot{V}_{O_2}$ is as small as Morrison found in his subjects. If the covariance found by Dwyer and Pilmanis[19] applies, then 2% inspired CO_2 could be reached in as little as 220 and 228 minutes.

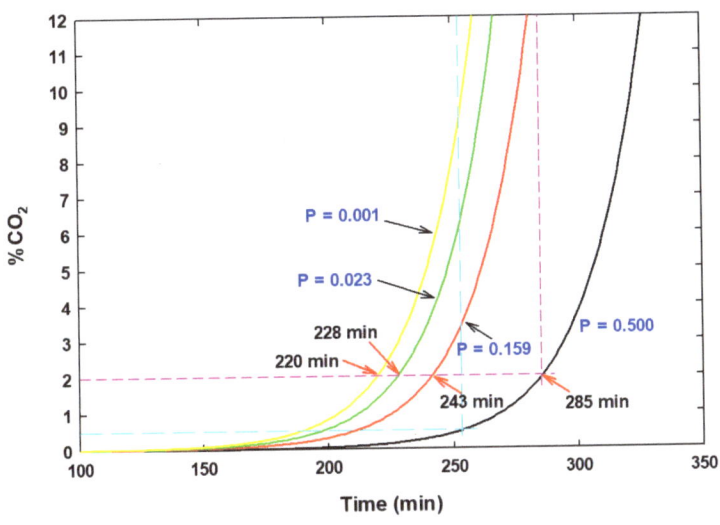

Figure 28. Breakthrough curves assuming a large covariance.

Figure 28 was based on data from Dwyer and Pilmanis[19], with covariance between R and $\dot{V}_{O_2} = 0.057$.

Table 3. Calculated time to reach 2% inspired CO_2 (Figures 27 and 28.)

	Probability			
COV	0.5	0.16	0.023	0.001
0.007 M	285 min	258 min	246 min	238 min
0.057 D&P	285 min	243 min	228 min	220 min

If a dive is conducted to the published mean breakthrough time of 253 minutes, then *by this analysis*, there is as much as a 16% chance that inspired CO_2 could approximate or exceed 2% CO_2.

Figure 29 emphasizes the probability of reaching very high CO_2 levels if dived by the clock to the time expected for breakthrough to 0.5%.

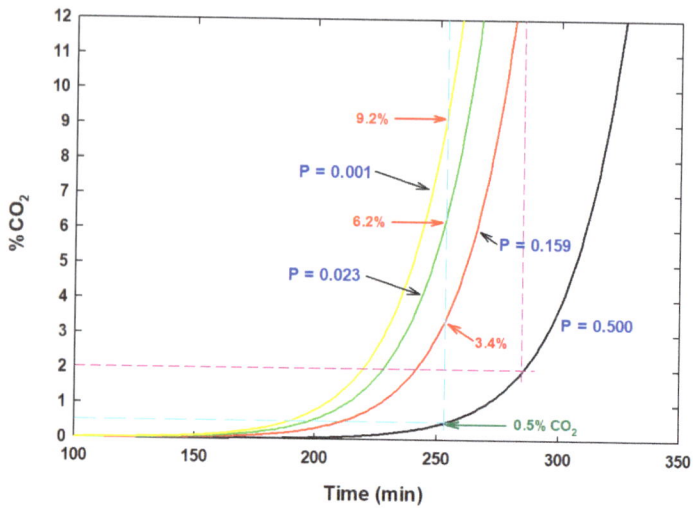

Figure 29. If dived to the anticipated breakthrough time, dangerously high inspired CO_2 could be encountered.

Figure 29 was again the Dwyer and Pilmanis[19] data with covariance between R and $\dot{V}_{O_2}$ = 0.057. Table 4 provides the respective probabilities when a canister is dived to the expected 0.5% mean CO_2 breakthrough time of 253 min.

Table 4. Percentage probability of elevated inspired CO_2.

COV	Probability			
	0.5	0.16	0.023	0.001
0.007 M	0.5%	1.7%	2.9%	4.1%
0.057 D&P	0.5%	3.4%	6.2%	9.2%

The curves and table are interpreted as follows: if 253 min is the mean time a canister would last before the CO_2 effluent reached a 0.5% CO_2 concentration, then the probability that a diver's canister would fail before that time is 0.5. That means 50% of the canisters would not last 253 min.

If a diver plans on a 253 min dive, there is a 50% chance of inhaling greater than 0.5% CO_2, regardless of what their physiological coefficient of variation is on that dive.

Even if the covariance is relatively low, say 0.007 as in Morrison's data, then there is a 0.16 probability that inspired CO_2 at the end of the 253 min dive would be 1.7% (intersection of the red curve with the vertical blue dashed line, tan cell in Table 2). There is a 2.3% probability it would be as high as 2.9% inhaled CO_2 for the same dive duration (intersection of the green curve with the vertical blue line, orange cell in Table 2).

Expressed as odds, there is an approximately 1 in 5 chance of inhaling almost 2% CO_2 and a 1 in 42 chance of inhaling almost 3% CO_2. There is a one in a thousand chance the CO_2 would be as high as 4.1% (yellow curve and red cell in Table 2).

However, if the diver's covariance on that dive day was as high as shown by Dwyer and Pilmanis, 0.057, then there is a 1 in 5 chance of inhaling 3.4% CO_2 and a 1 in 42 chance of inhaling 6.3% CO_2. Due to the high risk, all cells are colored red in the bottom row of Table 4.

If one were to reject the conventional definition of 0.5% CO_2 for canister breakthrough limit and instead opt for 2.0% as the limit, predictions based on the propagation of error would become dire. Even with a small COV, there is a 1 in 5 chance of inhaling 7% CO_2. A 7% inspired gas mixture would soon render a diver incapable of function, and a concentration of over 10% can lead to convulsions, coma, and death[22].

Table 5. Mean canister breakthrough at 2.0% CO_2 = 285 min.

COV	P			
	0.5	0.16	0.023	0.001
0.007 M	2.0%	7%	11.8%	>12%
0.057 D&P	2.0%	>12%	>12%	>12%

The propagation of error analysis in Table 5 reveals no reliable margin for error when choosing a breakthrough limit of 2.0% CO_2. The additional dive time of 31 minutes can come at an unacceptable cost.

Only one of the sources of variability was examined in this analysis. Hence, a diver's uncertainty about his inspired CO_2 level[n] when using a time-based dive plan is even greater than shown in this example.

Monte Carlo Analysis

An alternative to the propagation of error analysis is required when variances are not distributed normally. The Monte Carlo technique was coded using various versions of Mathcad by Mathsoft, Cambridge, MA, with eventual conversions to PTC Mathcad 15.0 and PTC Mathcad Prime 3.1.

Figures 30-32 are individual breakthrough curves with parameters randomly chosen from the distributions shown in Figures 22-25. The values of variables in Equation (6) were randomly selected from the known distributions of sodalime packing mass, oxygen consumption, R (or RER), and ventilation equivalent.

---

[n] There is a wealth of information on the effects of elevated CO_2. One of the more recent articles[23] is from 2017.

The Monte Carlo analysis for non-normally distributed parameters yielded qualitatively similar results to the normally distributed cases.

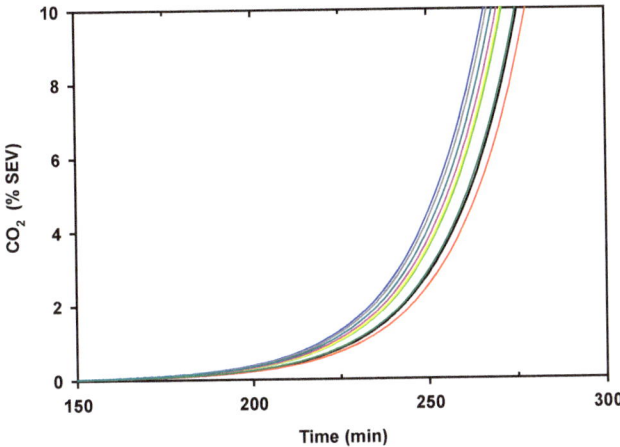

Figure 30. CO_2 curves with absorbent mass chosen randomly from the measured distribution.

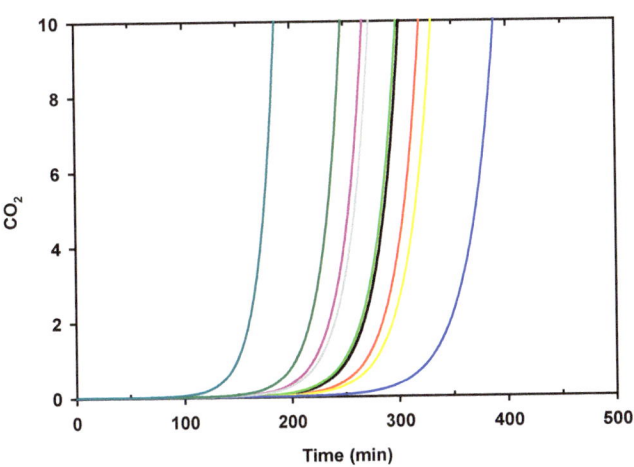

Figure 31. Combined variance in mass and $\dot{V}_{O_2}$.

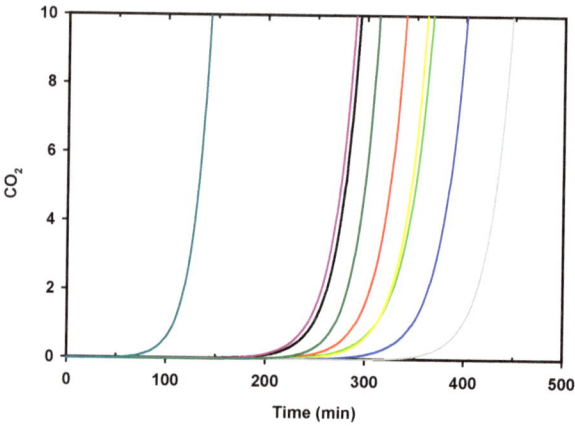

Figure 32. Combined variance in mass, $\dot{V}_{O_2}$, R and K_{O_2}.

Conclusion

Returning to this chapter's opening question: Are you "above average?"

If you're above average in $\dot{V}_{O_2}$, $\dot{V}_{CO_2}$, or $\dot{V}_E$, that does not bode well for your canister. If you are below average for temperature, you may also suffer premature canister breakthrough.

Being able to safely dive to a single published canister duration, such as 180 minutes is a fantasy. There are far too many variables that determine canister duration.

Planning your dive so you'll run out of oxygen before you run out of canister is a step in the right direction since oxygen consumption and carbon dioxide are related. However, that ignores the effect of water temperature on canister duration.

Ending your dive when you get cold is another step in the right direction, but if you add active heating to your thermal protection garments, your body may stay warm, but your canister won't.

Adding "conservatism" to a dive is not just for adjusting decompression tables. For years, the U.S. Navy has added conservatism to canister duration tables by making many unmanned "canister runs" over a wide range of temperatures. When the resulting durations are plotted, "prediction limits" define allowed canister limits. Those tables show that 97.5% of the experimentally determined canister durations lasted <u>longer</u> than the published limit.

The *prediction* is that such a result will continue into the future. (Is anyone taking bets?)

Even then, those results are only for a few measured temperatures. The durations for different temperatures are estimated through interpolation. The results also only apply to fixed rates of ventilation and CO_2 production. The odds of you exactly matching those parameters on your next dive is remote.

All these concerns argue for the use of some sort of CO_2 sensor. The most common of those in current use are thermal sensors that don't actually measure CO_2. They only respond to the heat patterns from active CO_2 absorption. In the next chapter, we will explore the complex thermal behavior of scrubber canisters. We will accomplish that through a new form of computer modeling.

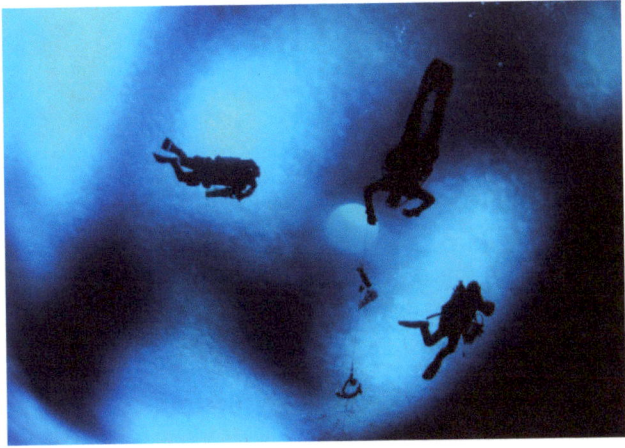

Rebreathers under the ice at McMurdo. Photo by Jeffrey Bozanic.

Underneath Antarctic ice, by Jeffrey Bozanic, Poseidon MK 6 CCRs.

Chapter 5. Computer Models

They say, "Seeing is believing."

When it comes to scrubber canisters, there isn't much to see, especially with absorbents lacking indicator dye[o]. So, this author created the computer simulations described in this chapter to allow divers to "see" inside virtual scrubber canisters.

Synchronicity

At the beginning of the 21st Century, two events happened that are pertinent to the following discussion. In 2001, I published the first account of what I now call the *Simulated* Physical Model, or alternatively, the *Stochastic* Physical Model, for scrubber canister thermokinetics[8]. Either way, the abbreviation SPM suffices.

A year later, Stephen Wolfram published his seminal book, *A New Kind of Science*[24]. The parallels between his thesis and my modeling approach were uncannily similar. In fact, in the opening pages of his gargantuan 1200-page tome, he made statements that parallel mine. Except, of course, his were much better worded.

---

[o] As indicated in Chapter one, the U.S. Navy banned indicator dye in absorbent due to the possible generation of toxic contaminates.

I was given permission from Wolfram Media to republish the following excerpt of *An Outline of Basic Ideas*, pages 1-2 of *A New Kind of Science*.

"If theoretical science is to be possible at all, then at some level the systems it studies must follow definite rules. Yet in the past throughout the exact sciences it has usually been assumed that these rules must be ones based on traditional mathematics. But the crucial realization that led me to develop the new kind of science in this book is that there is in fact no reason to think that systems like those we see in nature should follow only such traditional mathematical rules.

—Stephen Wolfram, "The Foundations for a New Kind of Science," in A New Kind of Science. (Champaign, IL: Wolfram Media, 2002), 1–2.

Ironically, it also took me twenty years to decide to publish the SPM model for a larger audience. Now let me state in my own words the approach described in this chapter and the rest of the book. The similarities to Wolfram's introduction will be less artful, but hopefully still apparent to the reader.

Like Wolfram, this author is confident that the universe is not based on mathematical equations. That may seem like a heretical statement, but I believe it is true. The universe as we know it operates on elementary principles that are not inherently mathematical. Instead, mathematics is the language humans use to describe the overall influence of those simple principles.

The most crucial physical principle for our application is that concentrations of heat and mass flow from an area of high concentration to areas of low concentration. Both heat and matter diffuse in that manner, expressing the universal tendency to maximize the entropy of a system.

Stochastic Simulated Physical Model

Devoid of the constraints of mathematics but controlled by randomness and probability, a stochastic SPM model can reproduce natural events far more complex than the usual mathematics, chaos and complexity theory excepted. Like any model, however, it is only helpful if it can predict experimentally observed events, especially unexpected ones.

Arguably, having gained some credibility by predicting anomalous events, as you will see in Chapter 6, the SPM stochastic model can reveal more detail than a deterministic, mathematically-based finite element model. That point will be made clear in the following pages.

Diving Application

In the SPM, we examine the probability of absorption reactions occurring within each small volume within a scrubber canister as affected by various internal and external factors.

To visualize events within an absorbent canister bed, this author developed a discrete computer model of CO_2 absorption in the scrubber canisters of closed-circuit underwater breathing apparatus (rebreathers.) The model was written in Microsoft Visual Basic 6.0 and was based on the repetitive application of small forces acting on up to 288,000 discrete volume elements (cells) within a simulated CO_2 absorbent bed in a rebreather scrubber canister.

The resulting heat and mass transfer influenced chemical reactivity in each cell. The warmer a cell, the greater the probability of an absorption reaction occurring due to increases in diffusivity of CO_2 and the increased energy available for chemical reactions.

Dynamics were determined by forces established at discrete time intervals and applied within discrete volume elements. CO_2 and heat moved according to gradients established at the preceding time interval. In Figure 33, the imagined diver's exhaled breath flows from left to right.

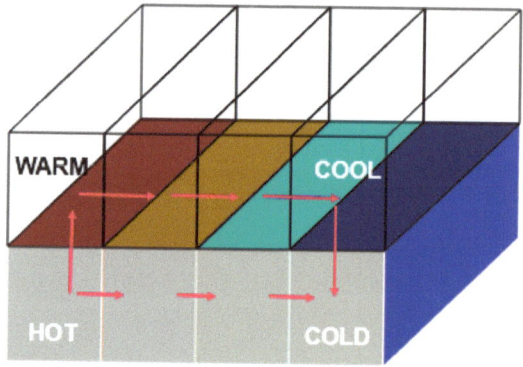

Figure 33. Heat flowed from hot to cold in the solid (grey) and gas phases.

The number of CO_2 absorption sites per cell depends upon absorbent granule size (Figure 34). Modeled granules were composed of varying numbers of concentric spherical shells and had random porosities and torturous water-filled channels running throughout. If an absorption site on the surface was unavailable, CO_2 molecules might diffuse into the interior (Figure 35.)

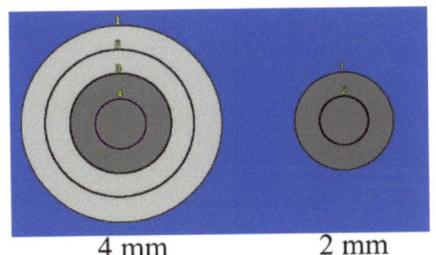

Figure 34. Two sizes of modeled spherical sodalime granules containing concentric spherical layers.

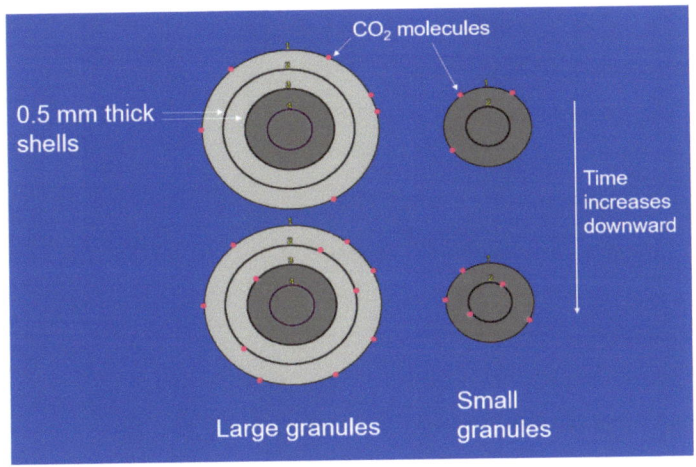

Figure 35. CO_2 molecules (red dots not to scale) encounter the granule's outer shell and, with time, diffuse into the inner shells.

Simulation Setup

The number of absorption sites per granule was directly dependent on the surface area of each granule shell.

4-mm shell = 32
3-mm shell = 18
2-mm shell = 8
1-mm shell = 2

Number of simulated reaction sites per granule (Figure 35)
4-mm granule = 32 + 18 + 8 + 2 = 60

2-mm granule = 8 + 2 = 10

<u>The number of granules per canister</u> was a function of granule size and absorbent bed porosity. For a bed porosity of 0.32, a 4.5-liter canister contains:
91,320 four-mm granules
730,500 two-mm granules

<u>Number of reaction sites per canister</u> = Number of granules x number of sites/granule
4-mm granule bed = 5,479,200 sites,
2-mm granule bed = 7,305,000 sites

Model Rules
- As a CO_2 molecule approaches a sodalime granule, it encounters a potential absorption site at random.
- If that site is empty, the probability that CO_2 will react is curvilinearly dependent upon temperature.
- As more absorption sites are filled, the fewer the available sites.
- A CO_2 molecule can loiter in a cell, randomly encountering absorption sites until its "residence time" for that cell has expired.
- If no free site has been found within the allotted cell residence time, the CO_2 molecule moves to the next cell downstream.
- Cell residence time is an inverse function of mass flow rate.
- Probabilistic rules determine the diffusion of CO_2 into the lower shells of a granule. The probabilities depend on: temperature, availability of an empty site, and diffusion barriers caused by accumulating reactants (calcium carbonate.)

Parameters (many set during program initiation: Figure 36.)
- size of the absorbent bed
- canister L / W ratio
- axial or radial canister
- canister insulation
- size of granules (2- or 4-mm dia.)
- water temperature
- initial bed temperature
- temperature dependence of reaction
- magnitude and temperature dependence of granule internal diffusion coefficient
- diffusion resistance of accumulating reaction products
- residence time
- heat of reaction
- thermal conductivity of absorbent granules
- specific heat of the gaseous phase
- CO_2 production rate
- probability of a volume element (cell) being filled.

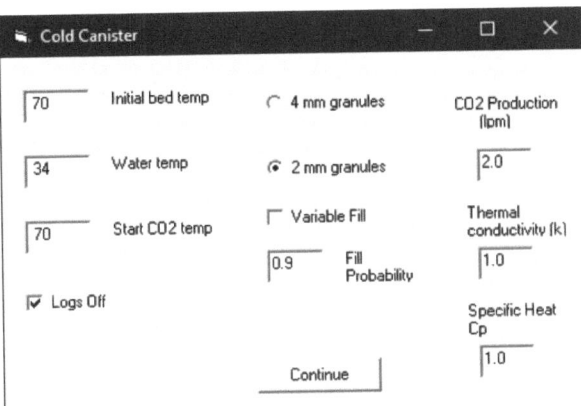

Figure 36. Canister Options

Model Outputs: Figure 37.

- temperature within each cell at any moment during the simulation
- amount of CO_2 stored in each cell
- *(Slicer Dicer* 3-D visualization software aided depiction of both temperature and CO_2 content data.)
- average temperature of the absorbent bed
- average temperature of the canister effluent.
- CO_2 released into the effluent per calculation cycle
- cumulative CO_2 in the canister effluent
- cells with free CO_2
- cells with absorption reactions

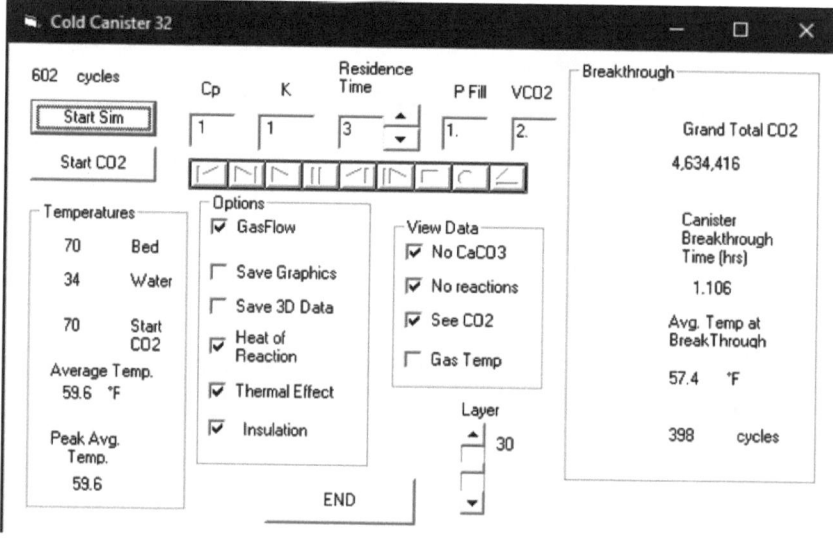

Figure 37. Simulation Results

As a reminder, CO_2 transport through the canister is divided into discrete time intervals, with absorption dynamics determined by forces established at each interval. Both CO_2 and heat from the exothermic absorption reaction are transferred to the gaseous and solid phase of each cell according to gradients established at the preceding time interval.

The probability that a simulated CO_2 molecule is absorbed in a particular cell depends upon the number of empty sites in the cell, the granule temperature, and residence time within the cell. CO_2 absorption reaction products create diffusion barriers within a granule, impeding intragranular transport. Absorption events, the total amount of CO_2 absorbed, and the temperature of both the gas and solid phase of each cell are depicted. Examples of these depictions are found in the following pages.

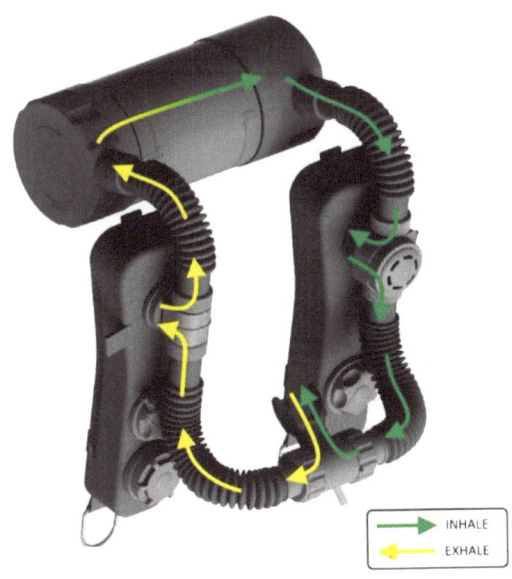

Figure 38. A typical rebreather gas circuit – the "loop," with the axial canister in the same orientation found below. Illustration courtesy of Dive Rite.

In Figure 38, used with permission from Dive Rite, the diver's exhaled gas containing CO_2 flows from the exhalation counterlung through the axial flow scrubber canister from left to right. The modeled gas flow path is shown in the figures below in the same orientation, with CO_2-containing gas flowing from left to right. With the exhalation counterlung surrounded by cold water, we assume that the temperature of the gas entering the canister is at the same temperature as the ambient water.

Based on rebreather temperatures in 28-29°F Antarctic water, data gathered by Heine and Bozanic[25] and reproduced by Menduno[26] (Figure 39), that assumption seems valid.

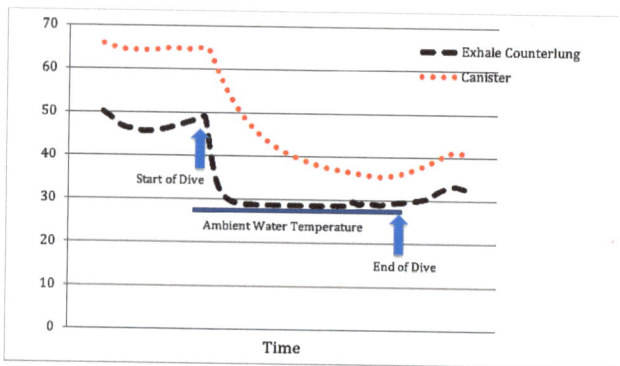

Figure 39. Data was gathered on an AP Diving Inspiration rebreather in McMurdo Sound. From references 25 and 26.

The time sequence views of the medial plane of the Z-axis are shown in Figure 40 in a small number of volume elements. Cells containing CO_2 reactions are green. Heat is indicated by red. Numbers represent computational cycle number (10, 100, 200, 300). With each computational cycle, CO_2-laden gas is deposited in each cell at the most proximal edge of the graphic (closest to the diver's exhalation hose, on the left side of each graphic).

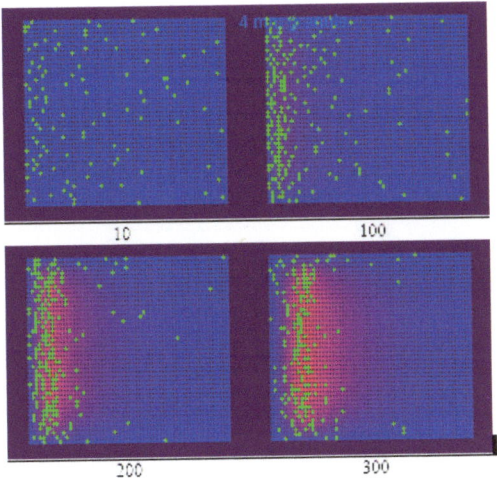

Figure 40. Snapshots for a time sequence of absorption are taken through a longitudinal section of the absorbent bed.

CO_2 (green square) enters from the left and exits from the right side of each canister cross-section. The x-axis is the axial distance down the absorbent bed. The number below each panel is the number of compute cycles when the image was captured.

Each graphic's upper and lower boundaries are chilled to the surrounding water temperature (34°F) and thus are blue. Exothermic CO_2 absorption reactions produce heat (various shades of red) which increases the probability of other CO_2 molecules being absorbed. That heat is spread granule to granule by conduction and is distributed downstream by convection.

Chemistry

The ability of sodalime to absorb CO_2 is due specifically to NaOH. The neutralization of CO_2 involves several chemical reactions[p], which can be summarized as the formation of the weak acid carbonic acid (Equation 8), its neutralization by the strong base, NaOH (Equation 9), with the evolution of heat. (Basically, any acid-base neutralization is associated with a negative enthalpy, meaning heat is produced through an exothermic reaction.)

(1) CO_2 (gas) + H_2O ↔ CO_2 (solution) (8)

(2) CO_2 (solution) + NaOH → $NaHCO_3$ (9)

(3) $NaHCO_3$ + $Ca(OH)_2$ → $CaCO_3$ + NaOH + H_2O (10)

The final step (10) results in the deposition of calcium carbonate ($CaCO_3$), and regeneration of the catalyst NaOH, along with the generation of additional heat.

Thermal Dependence of Reaction Probability

The probability of a chemical reaction, like that in Equations 8-10, is determined by the so-called *activation energy* of that reaction and the available energy; i.e., the temperature of the reactants. The well-known empirical Arrhenius equation describes the resulting rate of reaction (k), or the probability of that reaction occurring at any given temperature T, all else being equal.

$$k(T) = Ae^{\left(\frac{-E_a}{RT}\right)}$$

where E_a = activation energy, R is the ideal gas constant (8.314 J·mol⁻¹·K⁻¹), and T is the temperature of the reactants in Kelvin.

---

[p] Rendell DJ, Clarke M, Evans M. *The effect of environmental conditions on the absorption of carbon dioxide using soda lime. In: SAE Technical Paper Series. International Conference on Environmental Systems. SAE Technical Paper, 2003.*

As previously mentioned, rebreather canister durations are affected by many factors. However, one of the most important is temperature. Due to the multiplicity of factors affecting the probability of a CO_2 absorption reaction, we cannot use the Arrhenius equation to find the likelihood of any given CO_2 molecule absorption reaction. However, we can justify using a hyperbolic equation *in the form* of the Arrhenius equation.

Figure 41 is a plot of the rate constant k versus temperature as described by the Arrhenius equation. As T increases, the numerical value of the exponential becomes less negative, thus increasing the value of k.

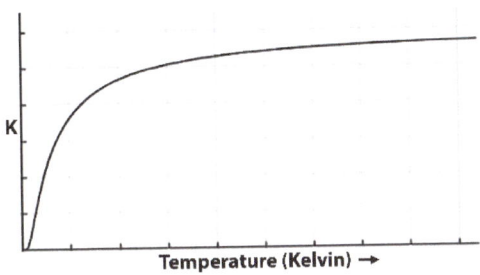

Figure 41. Arrhenius equation.

Figure 42 is the SPM-modeled relation between CO_2 absorption probability and individual cell temperature.

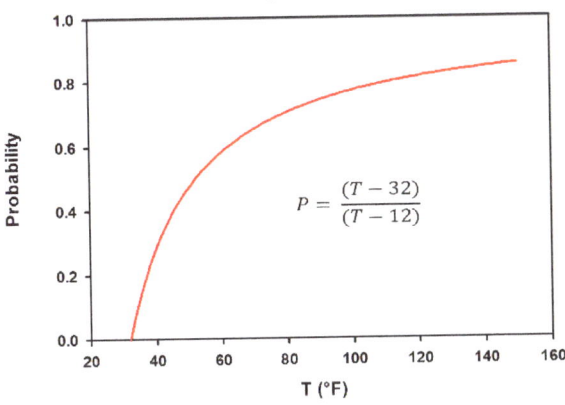

Figure 42. Absorption probability in the SPM model.

69

T is individual cell temperature at any given moment in time.

Measured parameters

Average canister temperature, individual cell temperatures, the average temperature of the canister effluent, and the number of CO_2 molecules exiting the canister are logged for each time interval. Separate models were created for axial flow and radial flow canisters. Since many transport coefficients were not known precisely, the SPM model was qualitative.

The large 3D data sets created from the simulation were visualized by *Slicer Dicer* software by Pixotec (Renton WA), formerly Visualogic. Some results are shown below.

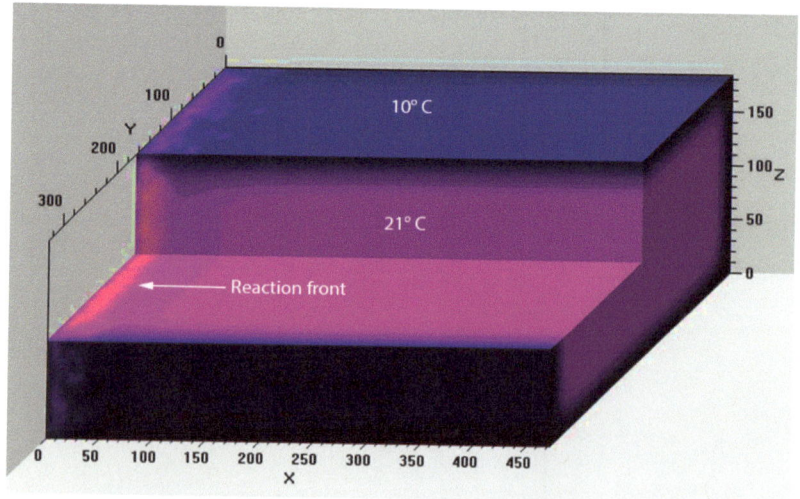

Figure 43. A warm CO_2 absorbent canister in cold water. Absorption reactions are beginning.

The X,Y,Z axes are the dimensions of the computational space in number of cells multiplied by eight for visual effect.

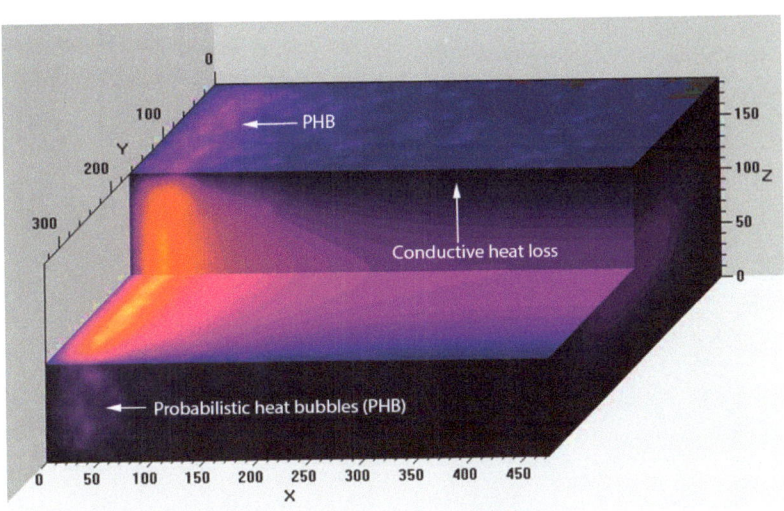

Figure 44. Heat is being lost to the water. PHB are Probabilistic Heat Bubbles.

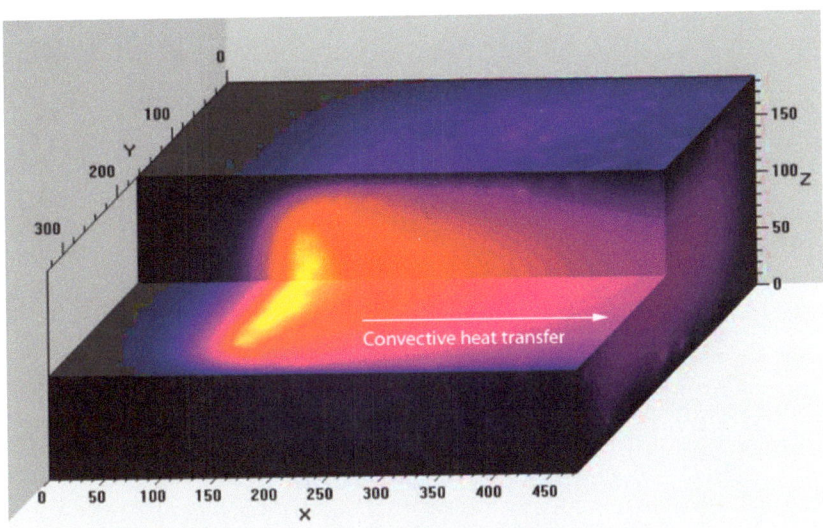

Figure 45. The heat of reaction is carried downstream, warming sodalime granules.

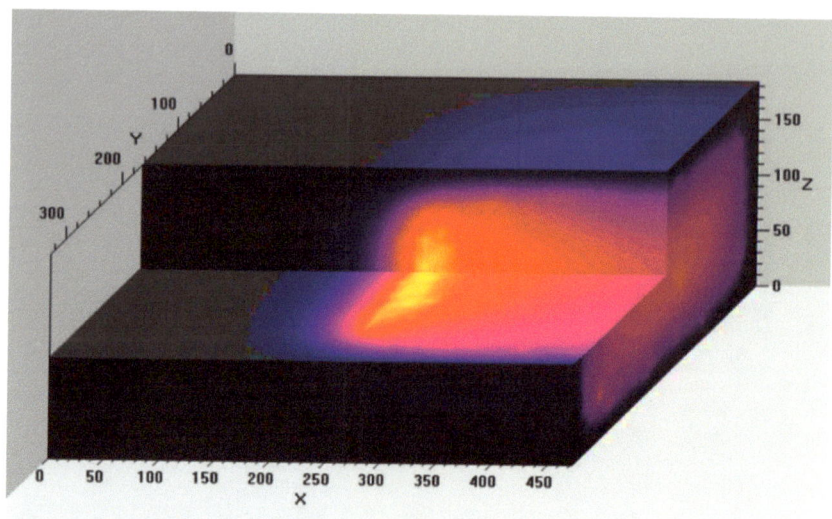

Figure 46. Halfway expended.

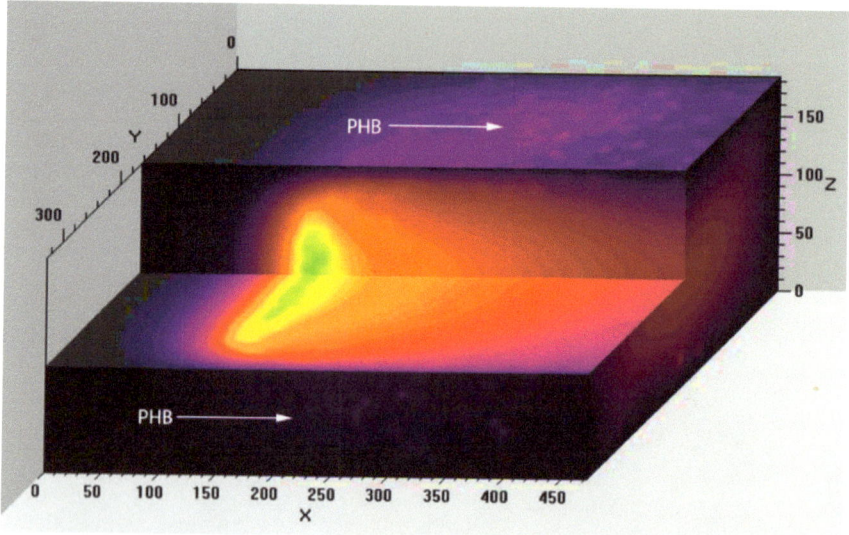

Figure 47. Color remapped to accentuate the stochastic nature of heat bubbling to the canister surface.

The inherent randomness of the stochastic model is revealed by the fine-structure bubbles (PHB) of warmth extending out to the periphery at the top and bottom of Figure 47. This image manipulated colors to increase contrast and make the heat bubbles more visible.

The probabilistic nature of the simulation is also apparent in Figure 48 when fluctuations in instantaneous CO_2 overflow (red tracings) are compared to the summed overflow (black exponential curve.)

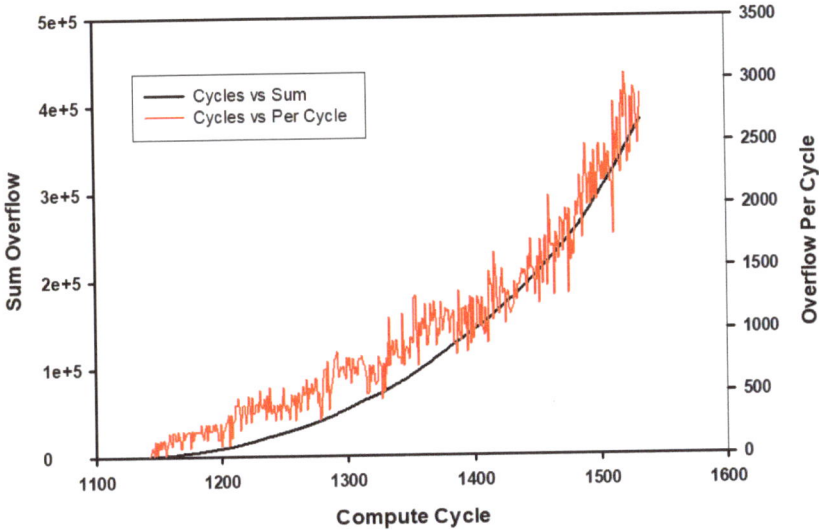

Figure 48. Fluctuations in CO_2 overflow counts per compute cycle and the sum of overflow events at a warmer temperature than the previous images.

The above traces were obtained from a simulation of 1.35 lpm $\dot{V}_{CO_2}$, in 70°F water and a residence time (Tr) of 3 cycles.

In Figure 49, cell temperature is plotted for two cells close to the proximal end of the simulated canister (the 5th cell from the entrance). Cells 5,6 lay near the edge of the canister, and thus temperature there was lower than in the middle of the canister, cell

5,12. "Cycles" are the number of elapsed computational cycles for the entire canister bed.

The fine-scale randomness of the stochastic process is revealed by the fluctuations of temperature with time at both volume elements (cells) being monitored.

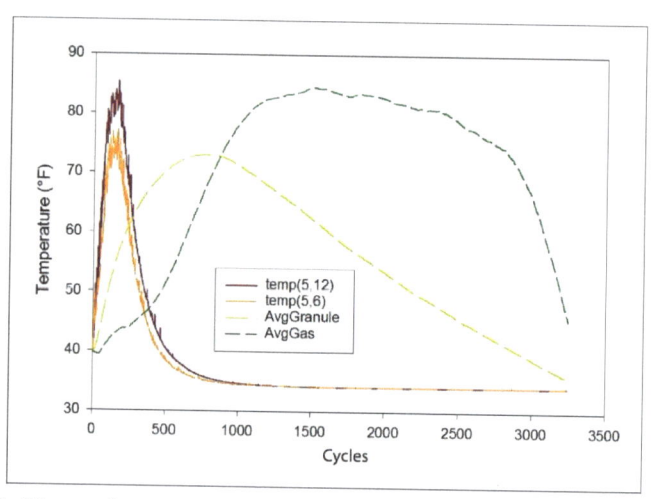

Figure 49. Plots of two specific cell temperatures near the entrance to the radial canister, and average gas and granule temperatures.

Although the average granule temperature (broken yellow-green curve) decreased almost linearly after reaching a peak value, the average gas temperature (dashed dark green curve) remained elevated until the canister was expended.

Graphs such as this, generated from the SPM, reveal the rationale behind thermal monitoring of scrubber canister activity with an NEDU-licensed temperature array used in various commercial rebreathers; e.g., the Sentinel, the Explorer, and rEvo, as well as the TempStik temperature monitor sold by A.P. Diving for their Inspiration rebreather.

Figure 50 is a similar thermokinetic SPM model as the others, but is specifically for a cylindrical axial scrubber canister as in

Figure 38. Although the computational space was still rectangular as in Figures 44-47, the absorbent reaction zones were constrained to a cylindrical shape.

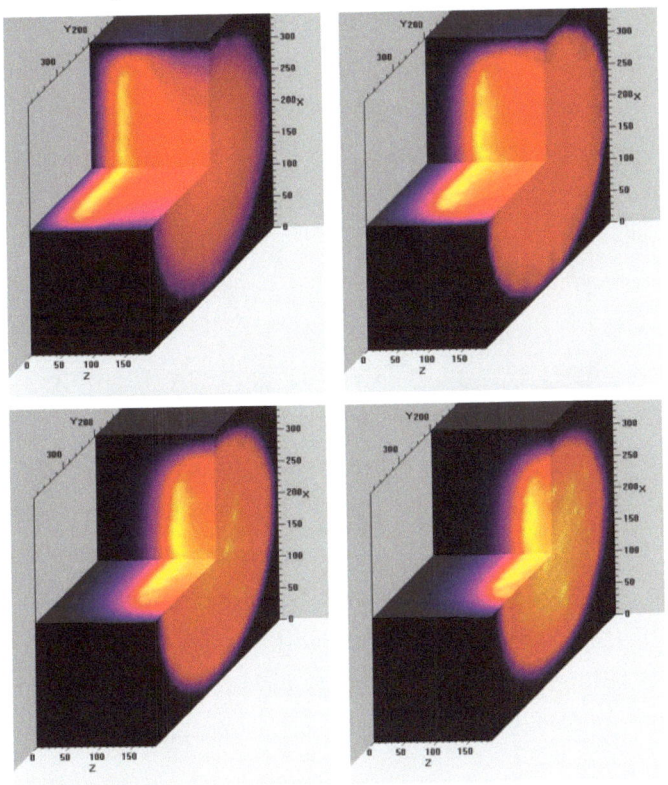

Figure 50. Simulation within a section of a cylindrical axial canister..

Decoupling Work and Ventilation

Compared to novice divers conducting a measured workload underwater, experienced divers are relaxed, conserving their air. Kerem et al[27] combined his test results with those of others to find that compared to novice divers, experienced divers breathed 24% less, and end-tidal CO_2s (an estimate of alveolar ventilation) were 23% higher. While both novice divers and nondivers maintained end-tidal CO_2s at about 40 mmHg, experienced divers maintained on average a value of 48.5 mmHg.

Another group of divers take their ventilation control to an extreme. They are hypoventilators, or so-called CO_2 retainers[27-29].

In my experience, CO_2 retention is a result, not a cause. The cause can be a severe challenge to the respiratory system due to high gas density, high hydrostatic or elastic loading, or heavy workload. Some divers simply abhor the sensations associated with labored breathing.

There are also some brains with unusually low sensitivity to arterial CO_2. When that low sensitivity is coupled with someone who hates the sensation of labored breathing, the outcome can be a suppression of breathing, even to unconsciousness.

For example, one of my Navy diver test subjects at the Naval Medical Research Institute grossly hypoventilated when tested at 300 fsw on a device that provided elastic respiratory loading. He kept altering his breathing rhythm, which managed to defeat the auto-control circuitry of the breathing device. (*Clever diver*, he thought.) But in the process, his exhaled CO_2 ("end-tidal CO_2") kept rising to dangerous levels. I warned him repeatedly to quit screwing around and breathe.

Instead, he laughed and said he was "tricking" the machine loading his breathing. (Actually, he used more colorful language.) Immediately after, my technician monitoring the mass spectrometer readings on the misbehaving diver's end-tidal CO_2 warned me that it had reached 90 mmHg. I immediately called for an "Abort." Everyone listening on the comms (and the tape) heard that *Abort* command except for the errant diver. He was already unconscious and falling from the bike to the metal grating.

Figure 51. Physiological experiments at depth. Hyperbaric Medicine Department, Naval Medical Research Institute, Bethesda, MD.

Figure 51 shows a similar experimental setup for another study, with a different diver than the one mentioned above. Navy Diver Frank Stout on a bicycle ergometer controlled by Navy Corpsman Tom Brisse at the NMRI's Man-Rated Chamber Complex, about 1985.

We know that under-ventilating (hypoventilation) causes arterial CO_2 to rise, which can abruptly lead to unconsciousness. But what about the scrubber?

That may seem like a strange question, but it had to be investigated. In so doing, interesting things were learned about scrubbers.

The case examined here by the SPM is that the simulated absorbent bed and water were all at the same temperature, 40°F. Varying amounts of CO_2 were sent into the scrubber with arbitrary values of $\dot{V}_{CO_2}$ ranging from 0.010 to 2.0. Residence time, Tr, was invariant and fixed at 3.

Those simulated canisters were then run to breakthrough. At the lowest $\dot{V}_{CO_2}$, 0.01, 3-million CO_2 *molecules* were absorbed at breakthrough. The canister lasted for 2.5 hours. The canister broke through at the highest CO_2 flow in 0.38 hours, capturing 1.8-million *molecules*.

As CO_2 started flowing into the canister, the bed began to warm at rates roughly proportional to the CO_2 injection rate (Figure 52). The slope of the rise at a $\dot{V}_{CO_2}$ of 0.020 lpm was 1.3 times that at 0.010 lpm. The slope at 0.030 lpm was about 1.4 times that at 0.015 lpm.

However, those proportionalities did not continue at higher injection rates. The slope at an injection rate of 2.0 lpm is not 20 times higher than at 0.1 lpm. In other words, the increase in warming with additional CO_2 injection becomes nonlinear beyond a certain point.

The length of each line stops at the moment of breakthrough. The blue line ($\dot{V}_{CO_2}$ = 0.01) ends at 912 cycles, the green (0.02) at 289, magenta (0.1) at 177, and red (2.0) at 134. So, it is true that the more the injection rate, the faster the breakthrough.

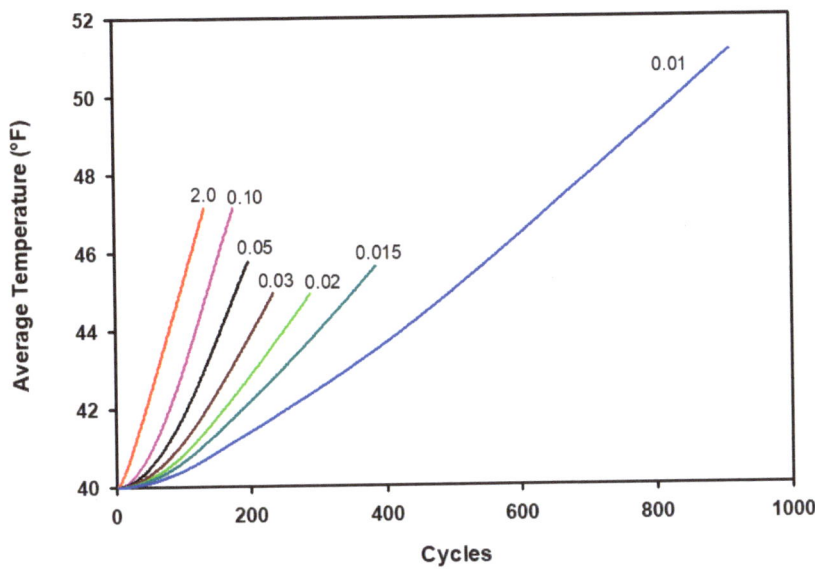

Figure 52. Average Temperature Over Time at Various CO_2 Injection Rates. $Tr = 3$.

However, Figure 53 reveals that there is a sharp transition in the $\dot{V}_{CO_2}$ versus time of breakthrough as CO_2 production rate increases beyond relatively low levels. The bottom panel in Figures 53 replots that data using the Weibull scale to expand the lower regions of CO_2 production. That scale is typically used in failure analysis, and seems reasonable in this case because breakthrough represents failure of the scrubber.

The transition from the vertical portion of the breakthrough curves to the horizontal portion is associated with consolidation of the exothermic absorption reactions.

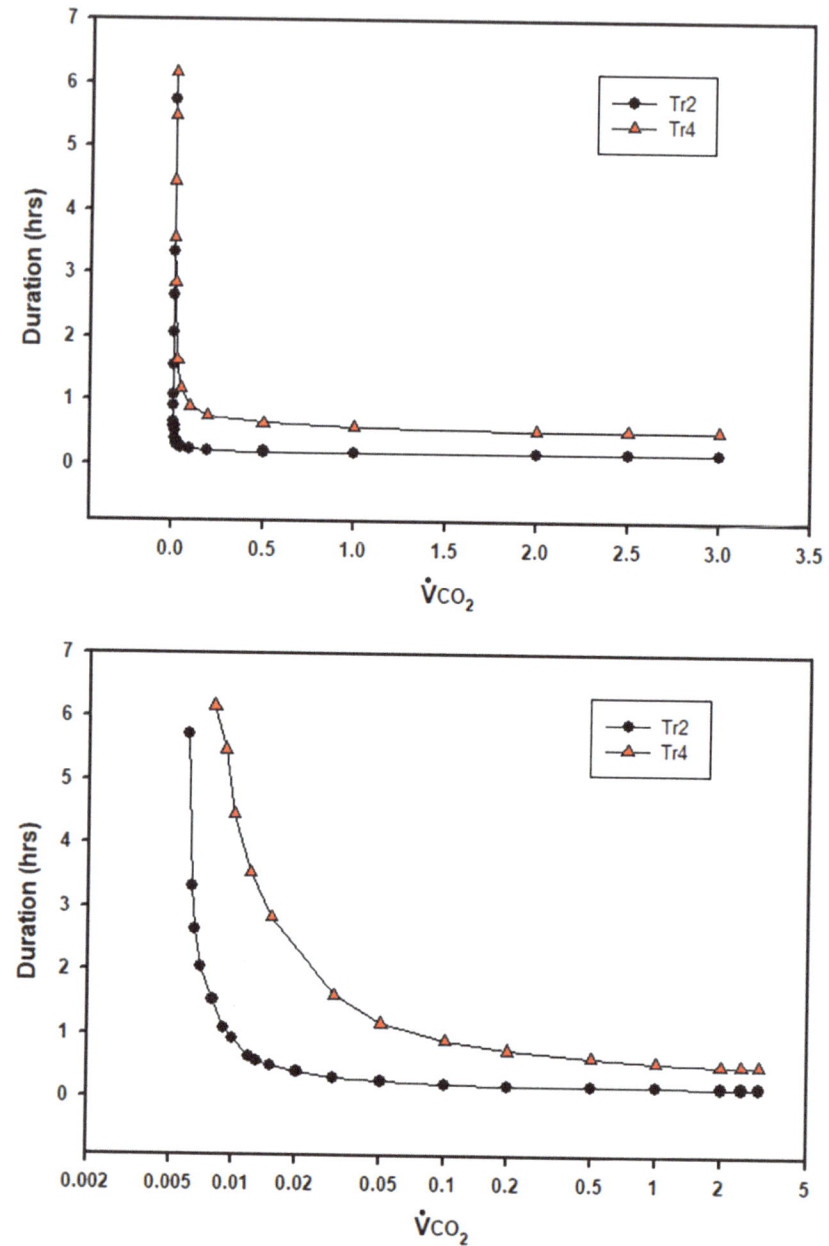

Figure 53. Breakthrough at 40°F as a function of CO_2 production rate and residence time (Tr).

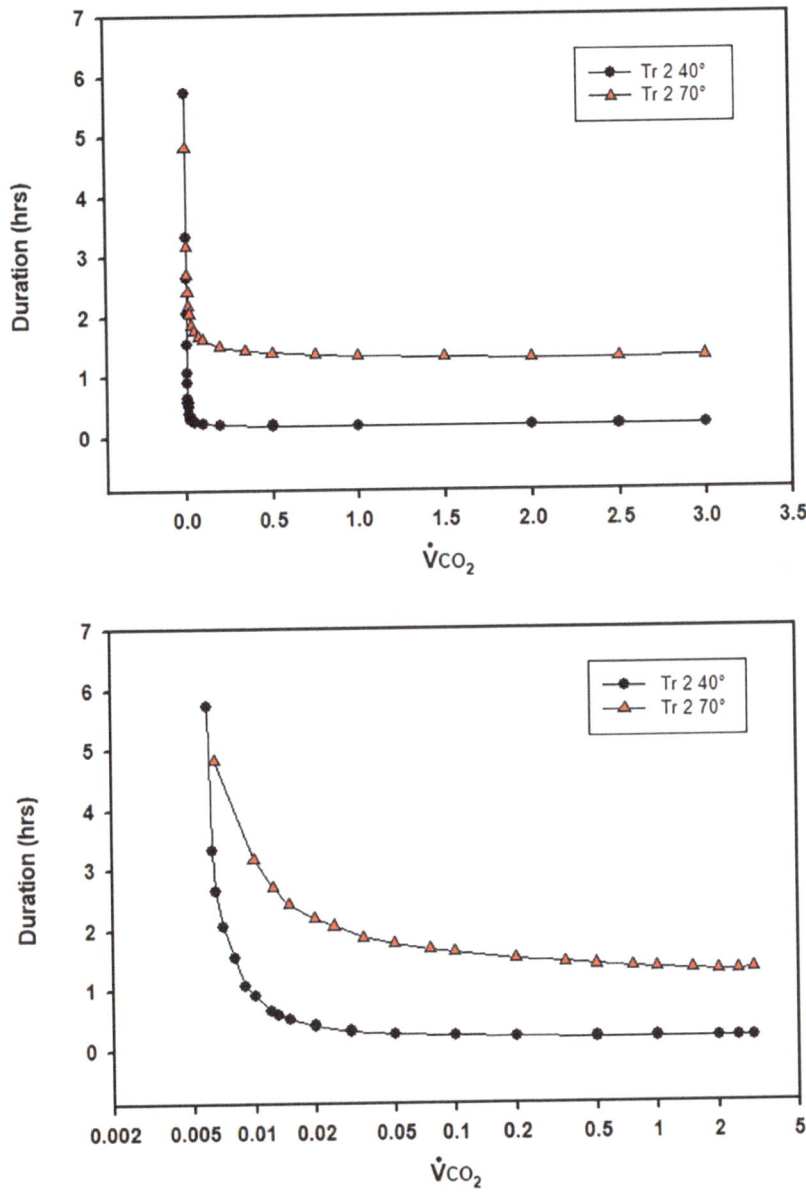

Figure 54. Breakthrough as a function of CO_2 production rate and temperature (40°F and 70°F.) All at $Tr = 2$.

At very low values of CO_2 production, CO_2 molecules within the canister remains scattered throughout, and warmer regions within the absorbent bed remain diffuse, as in Figure 55. (The number at the upper right is the simulation cycle number captured shortly after breakthrough.)

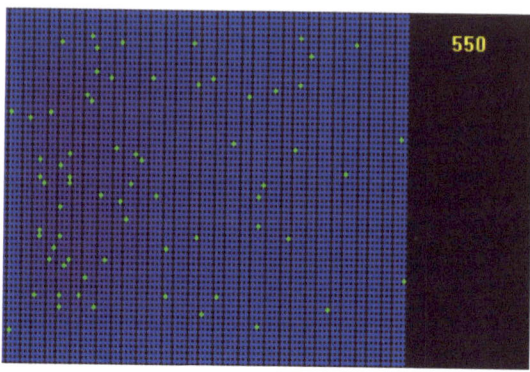

Figure 55. At breakthrough. $\dot{V}_{CO_2} = 0.015$.

In this condition, the slope of the duration versus $\dot{V}_{CO_2}$ curve is very steep. Duration is heavily dependent on $\dot{V}_{CO_2}$.

As $\dot{V}_{CO_2}$ increases, there is enough heat produced to accelerate the absorption reactions, causing consolidation of both the CO_2 molecules being absorbed, and the heat produced by that absorption (Figures 56 and 57.) Due to the Arrhenius-like nature of the absorption probability curve, a positive-feedback mechanism is established, producing heat which results in accelerated reactions producing still more heat. The only thing limiting the reaction is the loss of heat through conduction and diffusion.

In some respects, this positive-feedback, heat activation of the CO_2 absorption reaction is akin to the function of automobile catalytic converters. Heat from exhaust gas is required for a catalytic converter to begin working (250°-300°C)[30], and once activated, the catalytic reaction is exothermic; it produces even more heat.

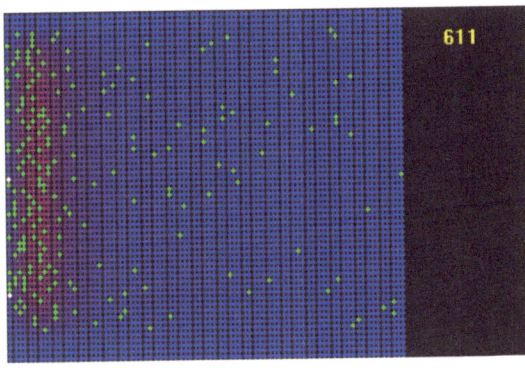

Figure 56. At breakthrough. $\dot{V}_{CO_2} = 0.1$.

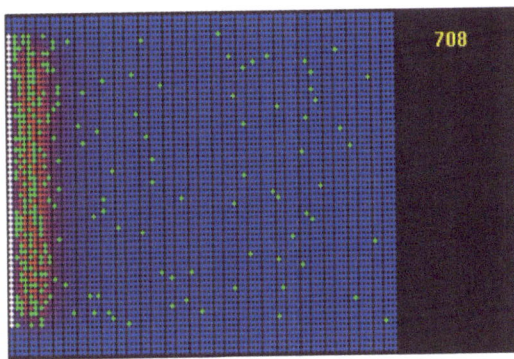

Figure 57. At breakthrough. $\dot{V}_{CO_2} = 2.0$.

Unlike catalytic converters, however, a rebreather's CO_2 scrubber is affected by the absorption. The actively absorbing region becomes used up, so to speak, and the consolidated region moves downstream. That downstream motion, however, is not paced by the amount of incoming CO_2, as in the pre-consolidation portion of the curves in Figures 53 and 54. Instead, it is determined by the residence time, Tr. That, in turn, is controlled by the average ventilation rate through the canister.

For that reason, the regions of $\dot{V}_{CO_2}$ above 0.1 in Figures 53 and 54 are mostly flat, with BT time being independent of $\dot{V}_{CO_2}$, but dependent on Tr.

It should be noted that thermal sensors like the TempStik rely on well-defined regions of consolidation. It is only in those regions of the canister bed that the thermal gradients are intense enough to be easily tracked. Ironically, according to the SPM model, those gradients are not determined by the rate of CO_2 inflow, but by how hard the diver is breathing.

As a reminder, physiological feedback mechanisms in the human brain keep breathing (ventilation) tightly locked with CO_2 production rate, i.e., work rate. That accomplishes the essential role of maintaining within normal limits the dissolved CO_2 concentrations in arterial blood.

It is only when a CO_2-retaining diver uncouples that compensation mechanism that untoward things happen. Keeping ventilation low in the face of a heavy workload will inevitably lead to headaches, CO_2 narcosis, and perhaps loss of consciousness, as happened in the Navy incident that began this discussion.

Ironically, a good thing to come from such an event is that the scrubber will benefit from a period of diver rest—if it doesn't flood.

Porosity

Imperfect canister filling leads to porosities not accounted for by the packing of perfect spheres. These porosities are distributed throughout the absorbent bed.

The imperfect packing of granules around canister walls leads to the so-called "Wall Effect" and "channeling." These sources and the effects of additional porosity are shown in Figures 58 and 59.

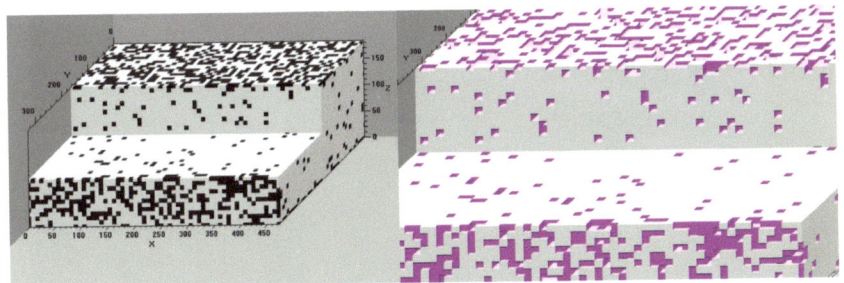

Figure 58. Edge and internal (colored) voids created probabilistically.

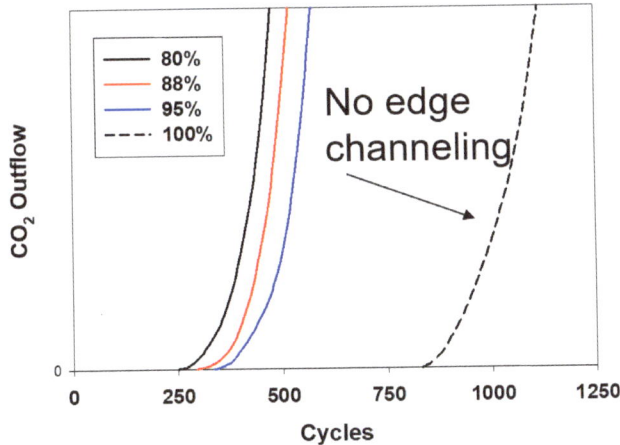

Figure 59. Effect of filling probability (in percentage) and edge channeling on canister breakthrough.

Canister Insulation

In the following five graphics, a granule thermal conductance[q] of 1.0 watts/(m·K) is assumed throughout an axial canister. The canister and its contents were considered at 70°F as it was plunged into 34°F water. The simulated diver worked moderately hard,

---

[q] *Actual granule conductivity is unknown. At 25°C, the thermal conductivity of limestone is 1.26-1.33 watts/(m·K) at 25°C, wet sand = 0.6 watts/(m·K).*

producing 2.0 lpm of CO_2 to be sent through the canister. All 3D data was collected after 450 computational cycles.

The previous graphics in this book have been created without an insulating layer surrounding the cylindrical canister. Such is the case for Figure 60, with a cross-section slice through the simulated canister.

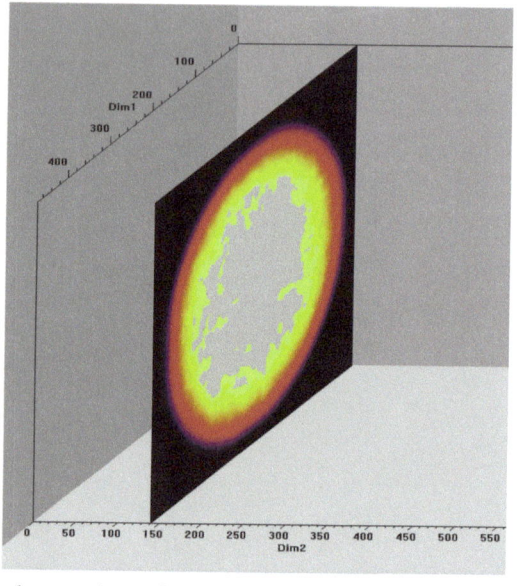

Figure 60. Thermal mapping of cylindrical canister cross-sections without an insulating canister shell.

In Figure 60, "Dim 1" and "Dim 2" are the cell numbers times 8. Once again, black in this thermal map represents 34°F, and white represents 130°F or above. Yellow, red, and purple represent progressively cooler regions.

This cross-section (about 20% down the long axis) was taken just after the reaction peak had passed to better define the random nature of the heating areas. The peak temperature was about 23% of the length of the longitudinal axis, so this cross-section had begun cooling down as the reaction center moved further right.

The low-contrast colors in the interior regions of the cylinder make the randomness of those temperature distributions hard to see. However, using a different color mapping, with the hottest area appearing black rather than white and the coolest shown as green, the randomness of the interior heat distribution is more visible.

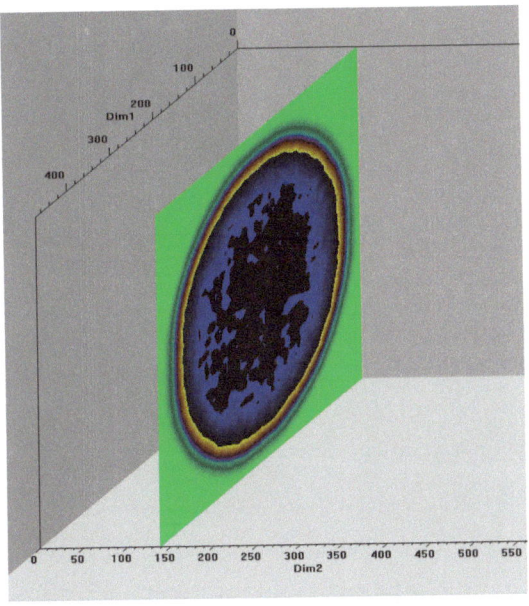

Figure 61. The same data as in Figure 60 except with a color map designed to differentiate between the hottest regions.

The simulated thermal imagery of Figures 60 and 61 does not appreciably change when comparing an uninsulated canister to an insulated canister. Instead, the thermal signature of an insulated canister is best seen by plotting the cell temperatures through the center of the canister, extending from one side of the cylindrical scrubber canister to the other.

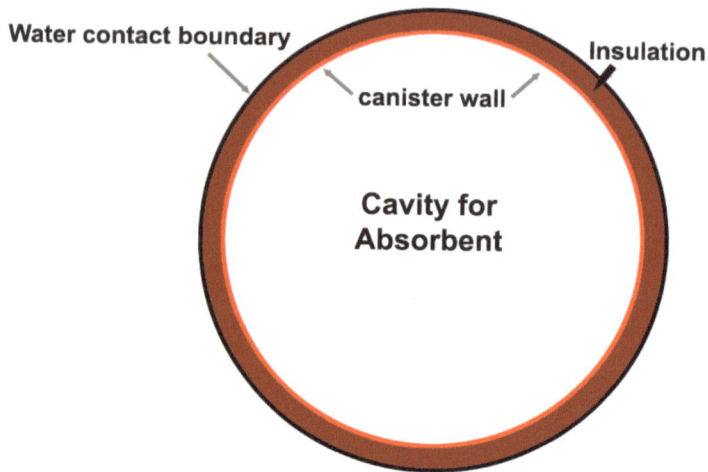

Figure 62. Insulating material (brown) surrounds the scrubber. Absorbent granules lie inside the canister wall (red circle).

Figure 62 illustrates the conceptual model for the next four simulations. There are two cylindrical boundaries affecting heat flow, one being the boundary between the absorbent bed and insulation, and one between the insulation layer and the surrounding water.

Figure 63 shows the simplest case of a canister filled with absorbent and equilibrated at 70°F. It is then dunked into 34°F water. Curves reveal temperatures through the middle of the canister at various distances down the canister. No CO_2 is flowing, so the plots are the two passive cooling curves after a time equivalent to 150 computation cycles. The red curve is the insulated[r] canister case, and the blue curve is without insulation.

---

[r] *A cylindrical insulating layer was created by establishing a cylindrical perimeter with thermal conductance of 0.01 watts/(m·K).*

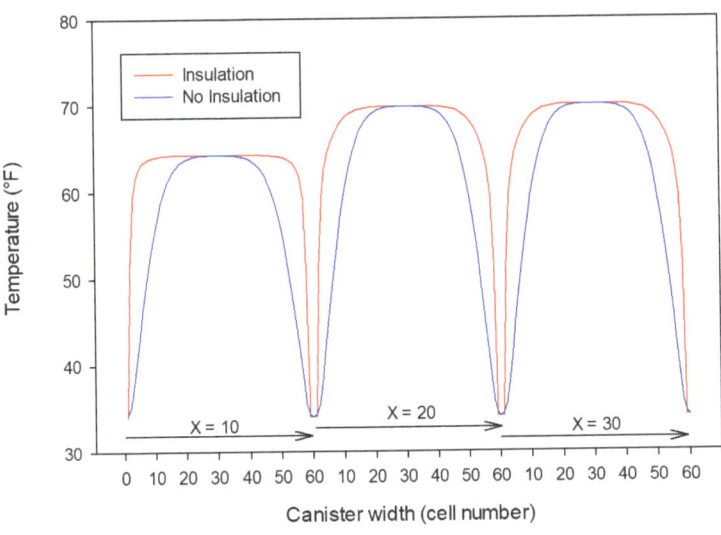

Figure 63. Insulation compared to no insulation during passive cooling of the absorbent bed, with no gas flow. Longitudinal lengths = 10, 20, and 30 cells out of a maximum of 80.

X represents distances down the canister. With an axial canister length simulated by 80 cells, x = 10 represents cross-sectional temperatures measured ⅛ of the way down the canister. X = 20 is the array of temperatures ¼ of the way down the canister. In the same manner, x = 30 is ⅜ of the canister length. The water temperature was 34°F, and the initial canister temperature was 70°F.

The large surface area exposed to low temperature at the opening of the canister quickly brings down the temperature of the cells closest to the opening (X = 0 to 10).

At the center of the canister, the temperature has not yet dropped from its initial 70°F. There is, of course, a significant difference in temperature at the insulated canister wall. Between those two regions, the absorbent is about 4°F warmer in the insulated case than in the non-insulated case.

In the next series of figures, temperatures are taken three distances down the canister as the diver's CO_2-laden exhaled breath flows through it.

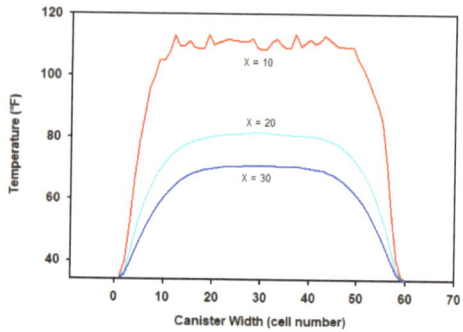

Figure 64. Uninsulated canister temperature profiles across three canister cross-sections after 150 computational cycles.

Figure 65 portrays the results for an insulated canister. The reaction front, and the highest temperatures, are located at $X = 10$. At $X = 20$ and 30, the absorbent granule cells have been preheated by the conduction of gas downstream from the reaction front. Since no absorption reactions are yet occurring at those regions, the temperature profile is smoothly curved downwards towards the edge of the canister. The canister cross-section at $X = 30$ is cooler than cross-section 20 since it is further from the active reaction front.

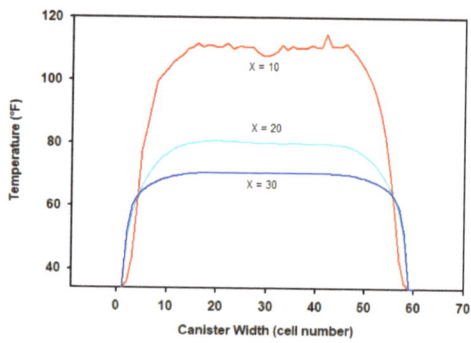

Figure 65. Same as Figure 64, but with insulated cylinder walls.

By definition, insulation has low thermal conductivity. The simulated insulating layer surrounding the granules has 1% of the conductivity of the granules and gas mixture within the canister. On average, the cooler volumetric cells were over 5°F warmer[s] in the insulated case than in the uninsulated case. That means the beneficial effects of the insulation were most noticeable in the cooler downstream regions of the canister (X = 20 and 30.)

After 450 computational cycles (Figure 66), the reaction front has moved a quarter of the way down the canister (X = 20), and the reaction region is fully involved, with maximal heat generated. The part around X = 30 (⅜ of the canister length) is heated by the hot gases flowing downstream from the reaction center. The area of X = 10 is cooling down since the reaction front has moved past it. However, it is still being heated randomly from both retained heat and continuing absorption reactions occurring within the deeper layers of the granules.

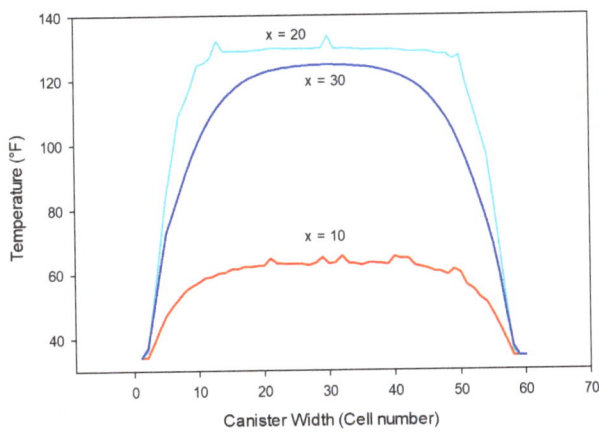

Figure 66. Temperature profiles after 450 computational cycles.

---

[s] *Those numbers were found by summing all the temperatures in a cross section, (e.g. blue or cyan curves) and dividing the sum by 60.*

As the previous graphics reveal, there is little difference between the modeled heat maps of the insulated and non-insulated canisters. Nevertheless, Figure 67 plots canister durations with water temperatures arranged across the known experimental diving range (28°F to 105°F). From 28°F to 40°F water, the canister duration was about an hour longer in the insulated canister than in the uninsulated canister.

In terms of simulated CO_2 molecules absorbed, at 34°F without an insulating layer, 2.5 million "molecules" were absorbed. With canister insulation, that number rose to 6.3 million.

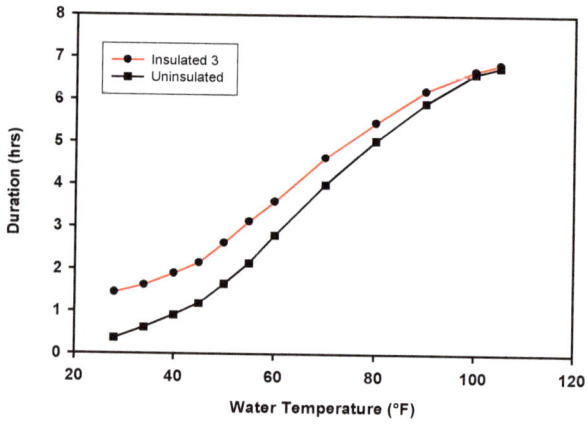

Figure 67. Canister duration as a function of water temperature for both an insulated (insulating layer K = 0.01 watts/m·K) and non-insulated canister (K = 1.0 watts/m·K).

As expected, the insulating layer is most useful at low temperatures.

Simulation Scaling

To put the simulation size into perspective, a large Micropore Dive Rite Optima cartridge is 124 mm in diameter (4.88 in) and 195 mm (7.68 in) long. With the simulation canister diameter being 60 units, the simulation canister length is proportional to 6.51 inches, not 7.68 inches.

Since computer memory is a limitation, the 288,000-cell canister model is 15% shorter than an actual Optima cartridge. However, the simulated canister length is certainly long enough to illustrate the essential features of canister function. Computed canister breakthrough occurs before the reaction front reaches the extreme end of the modeled canister, even at the hottest water temperatures (105°) (See Appendix A.)

Nanoscale variability

When running a stochastic model, you depend on a degree of randomness. In this model, random processes affect the access of a 0.33 nanometer long carbon dioxide molecule to a CO_2 absorption site on the first encountered sodalime granule. Secondly, probability and history will determine whether that site is available to accept a CO_2 molecule.

You can imagine a CO_2 molecule being engaged in a high-speed game of musical chairs. Except, any molecule not finding a "chair" within a compute cycle must move downstream to another ring of chairs, chairs which look suspiciously like another 2 or 4-mm wide sodalime granule. Depending on gas flow rate, temperature, residence time, and randomness, that CO_2 molecule will encounter more opportunities to be absorbed.

The result of the simulation on a macro scale is canister duration, taken as the time at which "canister breakthrough" is achieved. The precise definition of a *breakthrough* in this instance is as follows: The simulation rules called for a reasonable but arbitrary 600 overflow events per computational cycle. Once CO_2 *molecules* escape the canister without being absorbed, an overflow count for each cycle is printed on the canister window, as in Figure 55. When that number meets or exceeds 600 per computer cycle, canister breakthrough is said to occur.

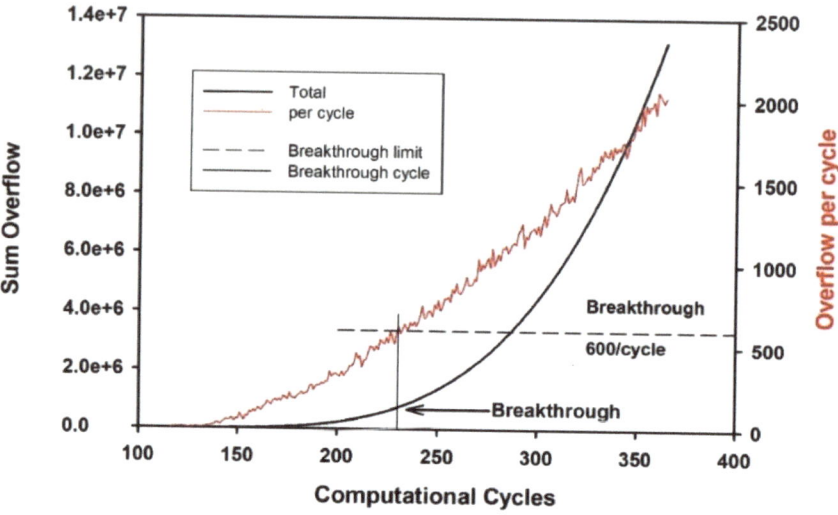

Figure 68. Definition of simulation breakthrough, an overflow of 600 "molecules" per computation cycle. In this example, breakthrough occurred after 230 computational cycles.

In Figure 68, CO_2 overflow per computational cycle and the total sum of CO_2 overflowing the canister is plotted for a 70°F absorbent bed placed into 34°F water. $\dot{V}_{CO_2}$ was 2.0 lpm. In this case, breakthrough occurred after 230 computational cycles.

Although a *breakthrough state* is triggered at the first instance of an overflow of 600 "molecules" per computational cycle, the randomness of the process means that the next computational cycle may show less than or more than 600 overflow events per cycle. That is reason for the ragged shape of the red tracing in Figure 68.

Also, screen prints of simulation windows are printed manually as soon after initial breakthrough as the operator's reaction time will allow. So again, the number recorded on the screen print may be scattered around a mean of 600 overflow events.

In a real canister testing laboratory such as NEDU, *breakthrough* is typically defined as the point where the partial pressure of CO_2 leaving the canister rises to 0.5 kPa. The stochastic canister simulation counts the actual numbers of *molecules* escaping the canister. The number of simulated CO_2 molecules absorbed within the bed at breakthrough can also be measured.

Table 6 describes canister breakthrough for one simulation run with a canister starting at 70°F and being dived in 34°F (1.1 °C) water. For this run, the gas circulating in the breathing loop was an Nx20[t] mixture with a constant pressure heat capacity of 1 J/(g·K), a thermal conductance of 1 watt/(m·K) with a canister 100% full of 2-mm equally sized spherical granules.

Table 6. Simulation results at breakthrough in 34°F water.

Tinitial	Twater	Dur (hrs)	# absorbed	cycles	Tavg °F
70	34	1.081	4.5e6	389	57.3

In Table 7, the initial conditions were the same, but the rebreather canister was dived in 70°F (21.1°C) water. The previous 1-hour duration in cold water was extended to almost 4-hours. The simulation ran about 3.5 times longer in the warmer temperature before canister breakthrough[u].

Caution: This is a model-specific simulation result, not a real-world result. Do not plan your dive based on a simulation!

Table 7. Simulation results at breakthrough in 70°F water.

Tinitial	Twater	Dur (hrs)	# absorbed	cycles	Tavg °F
70	70	3.822	12.74e6	1376	82.8

---

[t] Nx20 = 20% O_2, balance N_2.

[u] *Using a modest home computer with 12 GB RAM and a 4 GB display card, the simulation runs at 55 cycles per minute. A run of 1376 cycles takes 25 minutes to complete.*

While most divers are only interested in canister duration, it is of note that in this simulator, the number of simulated CO_2 molecules absorbed ranges from about 5 million to 21 million molecules.

Certainly, the number of molecules absorbed in an actual scrubber canister is unknowable but far greater than in the above tables. For instance, if one mole of CO_2 is absorbed, then under "standard" conditions (STPD), there would be $6.022 \cdot 10^{23}$ (Avogadro's number) of molecules absorbed.

For a diver working hard enough to consume 1.5 lpm of oxygen and produce 1.35 lpm of CO_2, a 2-hour dive would generate 162 L CO_2 at standard (STPD) conditions. According to the perfect gas law, one mole of CO_2 at STPD equals 22.4 L of gas. Therefore, that 2-hour dive produces 7.23 moles of CO_2 at standard conditions. Since one mole contains $6.022 \cdot 10^{23}$ molecules, a CO_2 scrubber would be tasked with absorbing $43.5 \cdot 10^{23}$ CO_2 molecules.

For the sake of time and computer memory, the existing canister model is limited in scope. Therefore, its results are not to be taken too seriously. Reliable real-world results can only be gained through tightly controlled manned testing, with its attendant wide variability, or through less variable unmanned testing such as that shown in Figure 11. However, despite the limitations of the SPM, it does illustrate the principles at play in the larger physical scrubber canister.

If the canister simulation is run repeatedly with the same initial conditions, the simulated canister duration will increase curvilinearly with temperature (Figure 69.) Each temperature shown, 28°, 34°, 50° and 70°F, reveals an average computed duration and a spread in those durations. That spread is indicated by one standard deviation bars on each side of the mean (or average) duration.

For low temperatures, the standard deviation is negligibly small, making the bars difficult to see. At higher temperatures, the deviation bars are noticeably wider. Presumably, that is because as time increases, the effects of randomness on the nanoscale begin to add up. Eventually, they become noticeable on the macroscale.

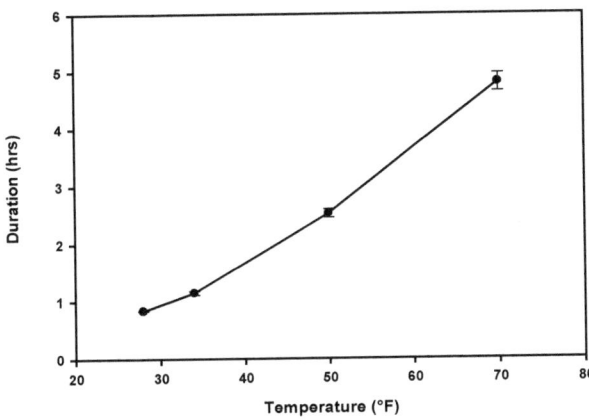

Figure 69. Means and standard deviation for five SPM runs under identical initial conditions.

Macroscale variability

Despite a technician's best effort to control variability in a testing laboratory, it is virtually impossible to eliminate large-scale or macro variability. For instance, when filling a canister with granular absorbent, it can never be guaranteed that every space within the canister is filled to the same degree, with no unexpected voids. There are also variances within the size distributions of the sodalime granules, and within those granules, variances in their absorbency.

Within the SPM model, it is possible to simulate variance in the probability of complete filling of the canister. Figure 70 is the mean canister duration (a black square in the 100% column) after five repetitive runs of a 70°F canister in 70°F water. The black circles represent the associated data points; tick label 1 being for the 100% filled case, and tick label 2 being the <100% filled case.

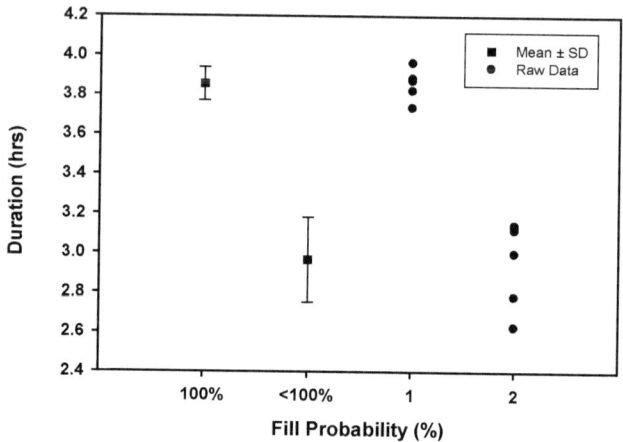

Figure 70. Using the SPM to test the effect of macroscale variability.

In the <100% column is the mean and standard deviation of six duration runs with 99, 98, 97, 95, 93, and 90% filling probability. (*The durations for the top three filling probabilities were partially overlapping.*) Since an increased chance of incomplete filling reduces the amount of absorbent available, the mean duration is reduced compared to the 100% filling case. Due to the range of filling probabilities within the sample, the resulting standard deviation is increased over that in the complete filling case. The standard deviation in the best possible case was 0.08 hours (5 min), while the standard deviation in the more variable case was 0.22 hours (13 min).

This simulated experiment suggests that careful and complete canister filling may be more important than you think.

Reaction Initiation

The following images (Figure 71) were created as part of the stochastic simulation of scrubber canisters. They depict the random heat patterns generated at the inlet end of a cylindrical, axial canister.

The images show the state of the canister as heat-producing CO_2 absorption reactions are beginning (A), are fully developed (B), and are waning (C) as the reaction front moves downstream (behind the plane of the images.)

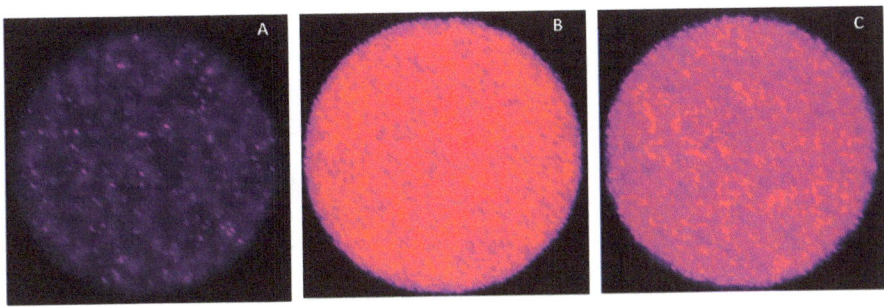

Figure 71. The inlet end of a simulated cylindrical axial scrubber canister during the time course of CO_2 absorption reactions. (Blue = cold, pink = warm due to exothermic reactions.)

The cooler (bluer) the proximal end of the canister (A), the less the amount of exhaled CO_2 reacting with the absorbent granules. The outer edges of the depicted canister slice are cooled by the low-temperature water surrounding the uninsulated canister.

Water Temperature

The effect of water temperature on the CO_2 breakthrough curve is shown in Figure 72. The Y-axis represents the number of simulated CO_2 molecules exiting the scrubber canister and being returned to the diver to be inhaled.

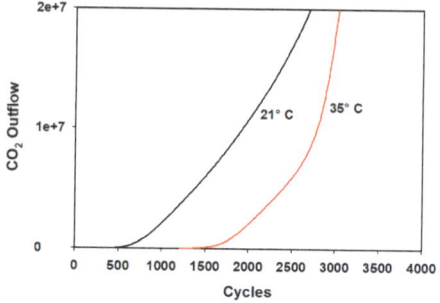

Figure 72. An example of the effect of water temperature on canister breakthrough.

Radial Canisters

The preceding models were based on an axial flow canister design, where CO_2-laden gas flows from one end of the canister to the other. Another canister design has radial flow, where gas moves from outside to in or vice versa (Figure 73). Some rebreathers accommodate either canister form factor.

Figure 73. One type of radial flow canister.

Exhaled gas enters the center channel of the absorbent bed and moves radially to an outer collection cavity.

In the following series of panels in Figure 74 (*numbered in sequence*) exhaled gas chilled by 34°F ambient water entered the canister, which had been previously equilibrated to 90°F. The cold gas entered a central circular channel, and then flowed from the central portion of the canister towards the outside. The modeled absorbent bed contained small grain (2 mm diameter) granules.

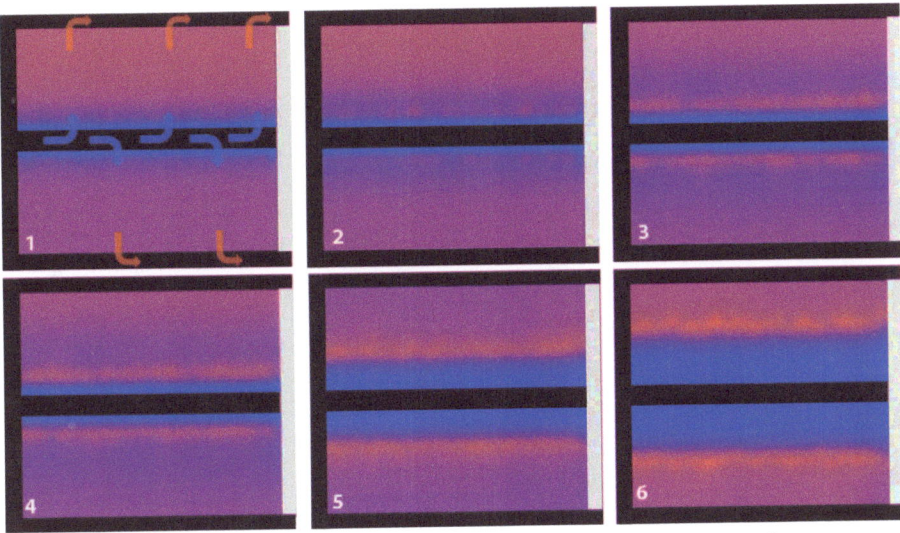

Figure 74. An exhaled breath flow path for an inside to outside-flow radial canister.

The colored panels in Figure 74 are cross-sections of an SPM model of a radial scrubber canister with water-chilled exhaled gas flowing from the inside to the outside of the canister in numbered sequence. The exothermic reaction front is clearly visible starting in frame 3. The center of the canister is chilled by convection until CO_2 absorption reactions begin heating granules. The canister core cools once the reaction front passes.

Flow Rate Dependency

CO_2 residence time within the canister (T_r) is controlled by the mean exhalation flow rate. The slower the diver's exhalation flow rate, the more time each CO_2 molecule has available to find an empty absorption site within the entirety of the absorption bed. When all the absorption sites have been filled, or if residence time is too short to allow a random walking molecule to locate an absorption site, the simulated CO_2 molecule exits the bed without being absorbed. Figure 75 plots the number of overflow molecules for T_r ranging from 2 to 4 arbitrary time units.

There are two primary features revealed by this simulation. As T_r increases from 2 to 3, the simulation time doubles before "breakthrough," defined as CO_2 overflow rising noticeably above zero. As T_r increases from 3 to 4, the breakthrough time almost doubles again. Secondly, the rate of rise of CO_2 overflow following breakthrough slows as T_r increases.

Since flow rate has a relatively large effect on "canister breakthrough" in the model, NEDU usually constrains flow to 40 lpm during its canister duration tests.

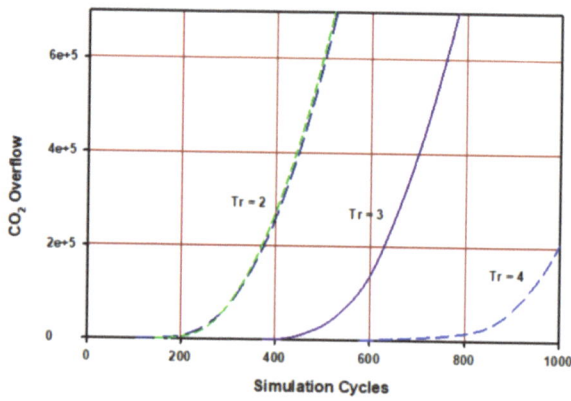

Figure 75. The effect of gas residence time on CO_2 breakthrough.

Real Data

For a reality check, this author performed forward stepwise regression on canister durations as a function of CO_2 injection rate and ventilation rate. Not surprisingly, the higher the CO_2 injection rate, the lower the canister duration, to a point (refer to the section on decoupling work and ventilation.) Conversely, the lower the ventilation rate, the longer CO_2 molecules reside within the absorbent canister ($T_r \uparrow$), and the higher the probability of them being absorbed on the first pass through the canister. In turn, that results in a longer canister duration.

However, the CO_2 injection rate and ventilation rate (and therefore Tr) are not independent. Both in breathing machine testing and physiology, the higher the injection rate, the higher the ventilation. Which translates to a shorter Tr.

The best regression to the duration data (in minutes) was:

$$Duration = Constant + (A \cdot \dot{V}_{CO_2}) + (B \cdot \dot{V}_E) \qquad (11)$$

where the coefficients A and B have units of min^2/L.

In the best fit, A = - 60.8 ± 38.3 (mean ± standard error of the estimate, SEE), and B = - 6.6 ± 1.9.

P values for parameters A and B were 0.022 and 0.014, respectively. (The value of *Constant* is not published for confidentiality reasons. However, a dummy constant is used below.)

The validity of the regression had been assured by the following pre-checks: both the normality and constant covariance test was passed, and the power of the statistical test with an α of 0.05 level of significance was 0.991 for a combined sample size of 20. The variable of *depth* was excluded from the model because it did not contribute to the fit.

For a dummy constant of 800 minutes, solving Equation (11) for a $\dot{V}_{CO_2}$ of 1.35 standard liters per minute and a ventilation of 40 lpm, we would project a duration of 454 minutes.

Increasing the $\dot{V}_{CO_2}$ to 1.6 lpm for the same ventilation, yields a projected duration of 439 min, a 15-minute reduction. Additionally increasing ventilation rate from 40 lpm to 50 lpm, reduces duration to 373 minutes, an 81-minute reduction.

Surprisingly, based on Equation (11), the reduction of the ventilatory flow rate through their testing canister with a <u>constant CO_2 injection rate</u> had more than double the effect on decreasing canister duration than did an increase in the CO_2 injection rate for a <u>constant ventilation rate</u>.

Figure 76 plots the SPM canister durations versus temperature for two CO_2 injection rates, with associated ventilation rates based on standard assumptions and their associated canister residence time (T_r).

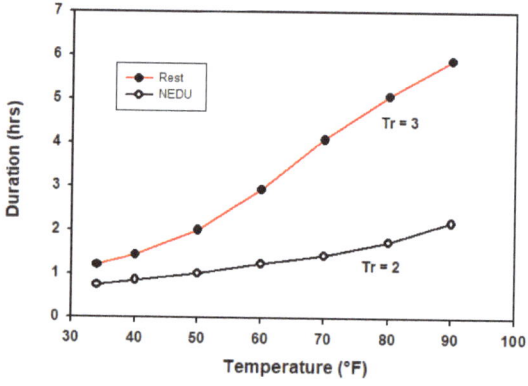

Figure 76. SPM canister durations under resting conditions ($\dot{V}_{CO_2} = 0.7$ lpm) and the NEDU protocol (1.35 lpm at 40 lpm.)

Based on the canister duration testing protocol used by NEDU, a 1.35 lpm CO_2 injection was simulated by a canister residence time of 2 ($T_r = 2.0$), and residence time at 0.7 lpm was set at 3 ($T_r = 3.0$).

Once again, we see that canister duration benefits more from increasing residence time than reducing $\dot{V}_{CO_2}$.

Carbonate Deposition

Figures 77 and 78 are SPM-generated images of the end-stage of the CO_2 absorption reaction, calcium carbonate deposition. It was obtained after 810 simulation cycles, with an initial bed temperature of 70°F and water temperature of 70°F.

Small grain sodalime (Figure 77) and large grain sodalime (Figure 78) are shown. The image colors reveal regions of complete carbonate deposition (yellow), partial (reddish tint) and beginning deposition (purple and blue tints). In this case, there was no canister insulation, so the canister walls were directly exposed to water.

In the bed of 2 mm diameter granules (Figure 77), breakthrough occurred at cycle 1293, 3.6 hr), and 18.696e6 "molecules" had been absorbed. With 4 mm granules (Figure 78), 14.184e6 molecules had been absorbed at breakthrough (1091 cycles, 3.0 hr) The ratio[v] of absorbed molecules (2 mm/4 mm) was 1.32, in line with the ratio of CO_2 absorption sites in the canister (1.33:1).

As before, in the following figures, axis labels of Dim1, Dim2, and Dim3 represent the number of simulation cells times some arbitrary multiplication factor that visually enhances the imagery. In all such graphics in this work that multiplicand is 8.

---

[v] *From Figure 35, there are 6 times more absorption sites on 4 mm granules than on 2 mm granules, but for a given size canister, there are 8 times more 2 mm granules than 4 mm. Total sites = 7,305,000 sites to 5,479,200 sites, 2 mm to 4 mm, respectively, for a ratio of 1.33/1.*

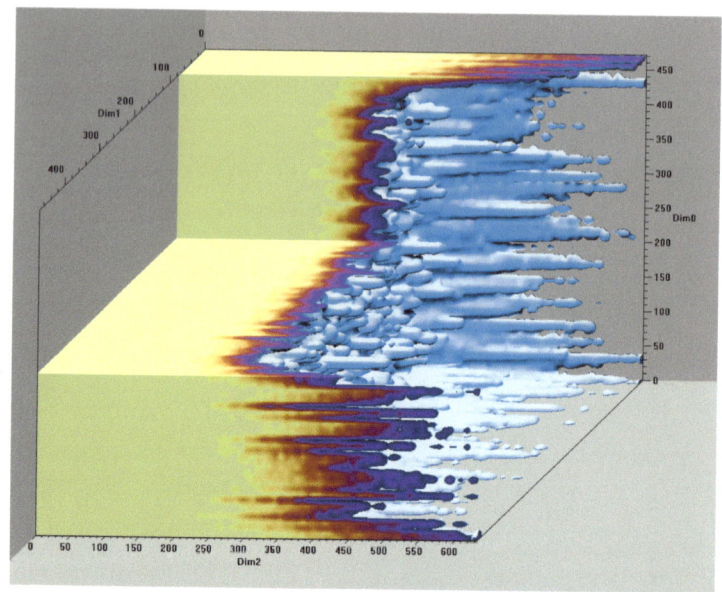

Figure 77. Carbonate in 70°F water at breakthrough. 2 mm granules.

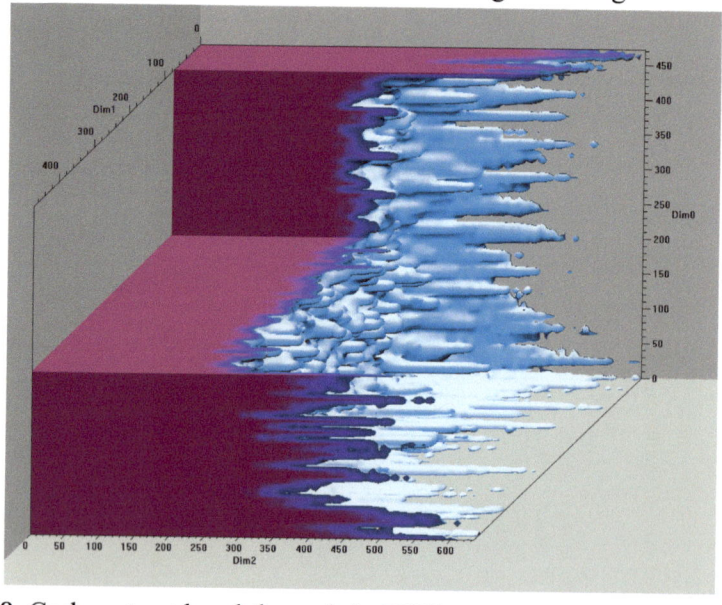

Figure 78. Carbonate at breakthrough in 70°F water. 4 mm granules.

The striking difference in color (yellow/tan versus reddish for the carbonate concentration was based on the SPM's account of CO_2

absorption reactions. (See Appendix A for details on color mapping.)

When the water temperature was lowered to 34°F, breakthrough times for 2 mm and 4 mm granules dropped precipitously (137 cycles, 23 min; 127 cycles, 21 min, respectively). Unlike in the case of 70° water, the ratio of molecules absorbed (2.150 million/2.074 million = 1.04) did not meet expectations based on the total number of CO_2 absorption sites.

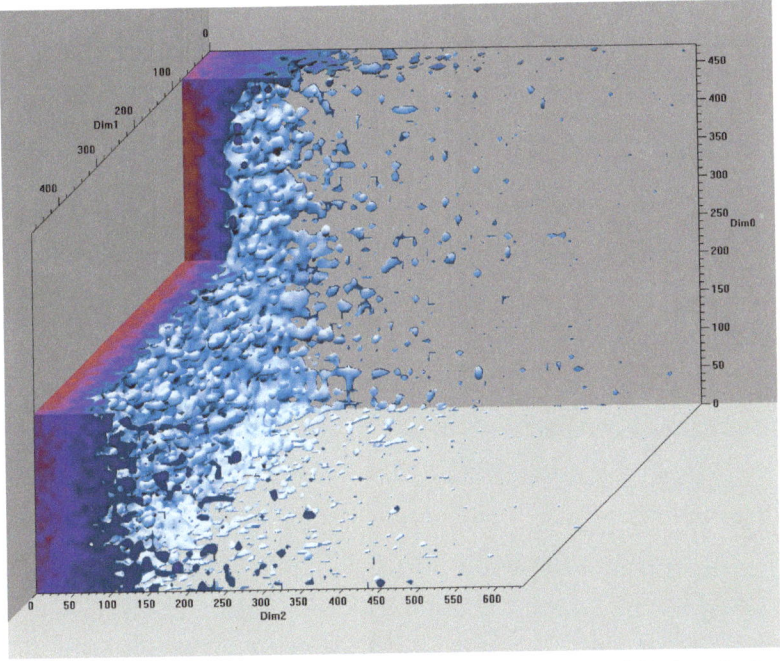

Figure 79. Breakthrough, 2 mm granules in 34°F water. No insulation.

The empty space in these figures is unused absorbent with no carbonate in the granules.

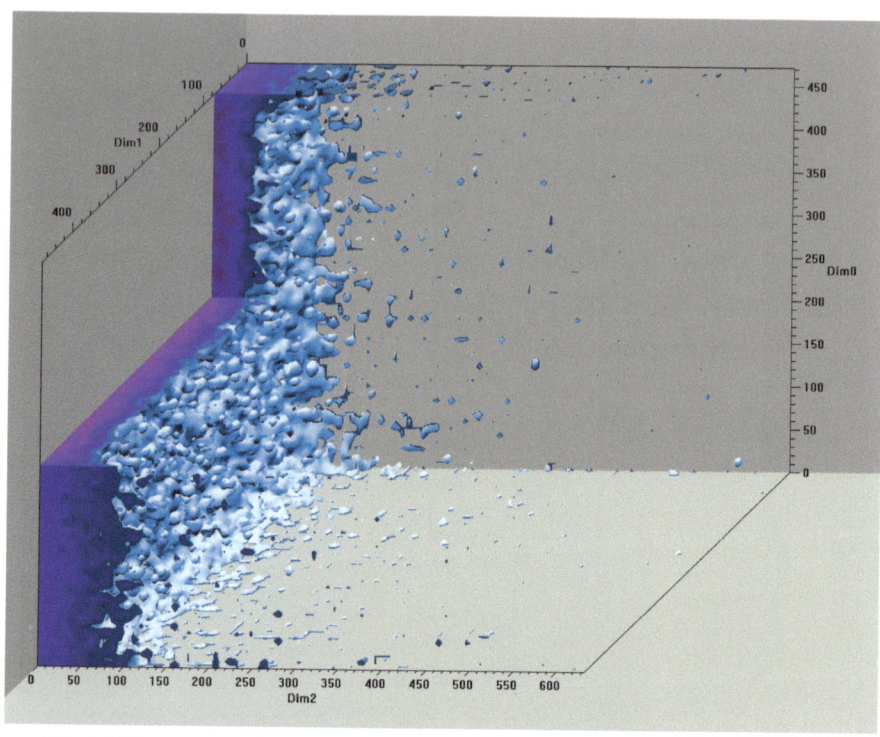

Figure 80. 34°F water and 4 mm granules at breakthrough.

To maintain consistency of the imagery with Figures 77 and 78, the following 34°F simulations were continued after breakthrough. The inspired gas would be unbreathable due to the high CO_2, but the simulation reveals interesting physics. In both Figures 81 and 82, the coating of the simulated canister walls with carbonate suggests that the low reaction probabilities along the cold boundaries of the canister allow CO_2 to penetrate deeply along the canister edges. Eventually, given enough opportunities, even low probability reaction events occur, causing carbonate deposition.

On the other hand, the high gas and granule temperatures in the middle of the canister favor rapid absorption of the incoming CO_2, slowing its penetration further into the canister.

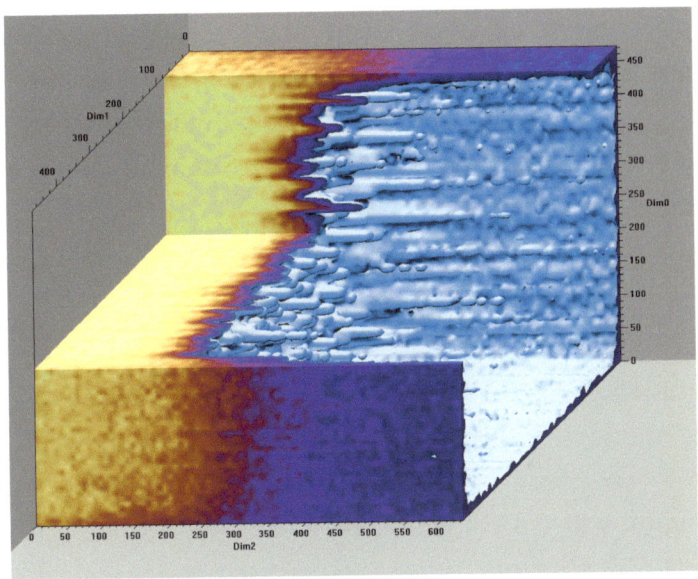

Figure 81. After breakthrough carbonate deposition, 2 mm granules, 70°F bed in 34° water without insulation.

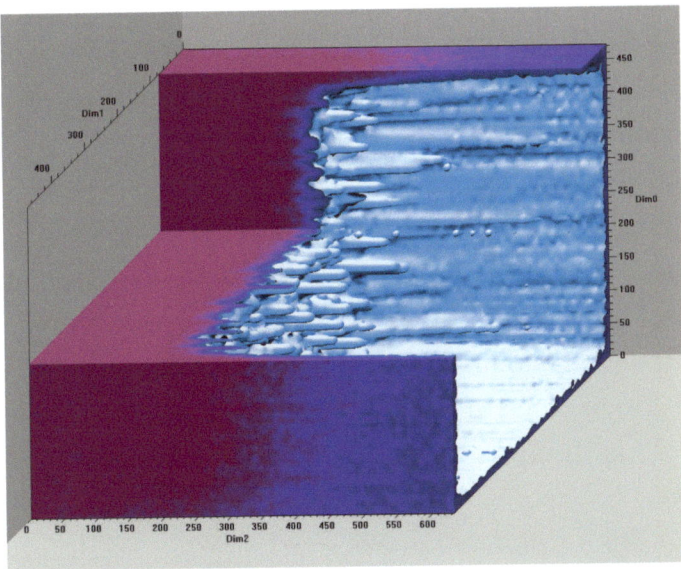

Figure 82. After breakthrough carbonate deposition, 4 mm granules, 70°F bed in 34° water without insulation.

As all rebreather divers know, spent absorbent inside canisters does not turn into a solid chunk of carbonate, despite what the imagery seems to show. The graphics simply reveal where sodalime granules with the greatest load of carbonate are located within the canister.

The Power of Simulation

Simulation is most interesting when it reveals something unexpected. The wall-coating phenomena revealed in Figures 81 and 82 was such a surprise. But to repeat, given enough opportunities, even low probability reaction events occur, causing carbonate deposition further down the rectangular absorbent canister than the main body of reactions.

Figure 53, showing a sharp transition from mass-based to flow-based canister durations, was another surprise. I had of late been puzzling over the lack of a strong dependence of durations on the amount of CO_2 entering the canister. Common sense says that dependency should exist.

It turns out that I was looking in all the wrong places. As I started simulating ever smaller $\dot{V}_{CO_2}$, I found what I was looking for. The more the CO_2 entering the canister, the shorter the duration.

I was taken aback, however, by the curiously sharp transition. But as it turned out, the word *transition* was the clue. The transition from a diffuse cloud of CO_2 and heat to a consolidated band of CO_2 and heat is strongly suggestive of something akin to thermal ignition.

In the study of forest fires, there is an oft-used phrase, *critical mass*. Once a mass of vaporized fuel reaches a critical level in the presence of heat, ignition and flames result[31].

Hopefully, no diver will ever have flames in their rebreather. However, the graphics of CO_2 absorption and the heat produced, paint a similar picture. At low CO_2 concentrations in cold canisters, the low probability of absorption causes CO_2 molecules to be widely dispersed. The heat from the occasional absorption reaction is diffusely spread among the absorbent granules.

As the mass of CO_2 molecules reaches a critical level, there is enough concentration of CO_2 at the entry to the canister to create a low-intensity version of a flame front. Like any flame, that chemical reaction front moves down the canister as reaction sites (fuel, if you will) become depleted.

At that point, the volumetric flow rate, and its attendant CO_2 residence time, becomes the rate-determining step. The lower the gas flow, the longer the residency time, and the higher the probability that the reaction front will dawdle, seeking out each available reaction site.

Fortunately for divers and the makers of the TempStik and other thermal monitoring systems, divers seem to produce enough CO_2 to generate an easily detectable reaction front. A diffuse zone of reaction would be hard to interpret by thermistor arrays.

According to Silvanius[32] et al., TempStiks work reasonably well at resting work rates (as during decompression) when tested at 19°C (66°F). However, there is currently no comparable data at 40°F (4.4°C), where the sharp transition in breakthrough time occurred (Figure 53).

Linear Flow

Figures 77 and 78 showing carbonate deposition, are two of the SPM images that illustrate the linear flow arrangement in this simulation. That linear flow is a desired feature used in Micropore

ExtendAir cartridges, absorbent cartridges which serve as a substitute for loose-fill absorbent granules. According to Micropore, straight-through flow, as shown in Figure 83, improves CO_2 absorption efficiency.

Figure 83 was redrawn and modified from a Micropore illustration of their ExtendAir cartridges.

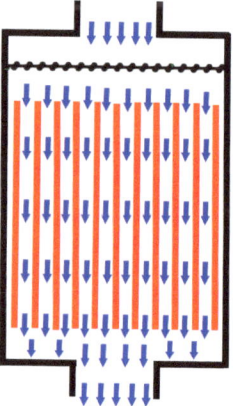

Figure 83. Linear flow in Micropore ExtendAir Cartridges.

Simulating the Impossible

CO_2 production and ventilation are irrevocably linked. Equation 11 shows that canister duration depends upon both $\dot{V}_{CO_2}$ and $\dot{V}_E$. Physiology demands that as physical workload increases, CO_2 production increases. However, to keep arterial CO_2 at physiologically tolerable levels, ventilation ($\dot{V}_E$) must increase to "blow-off" the extra CO_2. If that did not happen, a person would eventually lose consciousness from CO_2 toxicity.

With a simulation, however, you can break that linkage. In other words, you can simulate the effect on a scrubber canister of a high $\dot{V}_{CO_2}$ with a very low ventilation rate.

That simulation gets interesting when you simplify the initial conditions. The complexities of large thermal transients can be lessened by assuming the same high temperature for both the absorbent bed and surrounding water, without the benefit of insulation.

What makes this unrealistic exercise worthwhile is that the images in Figure 84 are suggestive of images in the literature. Figure 84 is a simulation run at 100°F for both absorbent bed and water, 2.0 lpm CO_2 production, and a residence time (Tr) of 4.0.

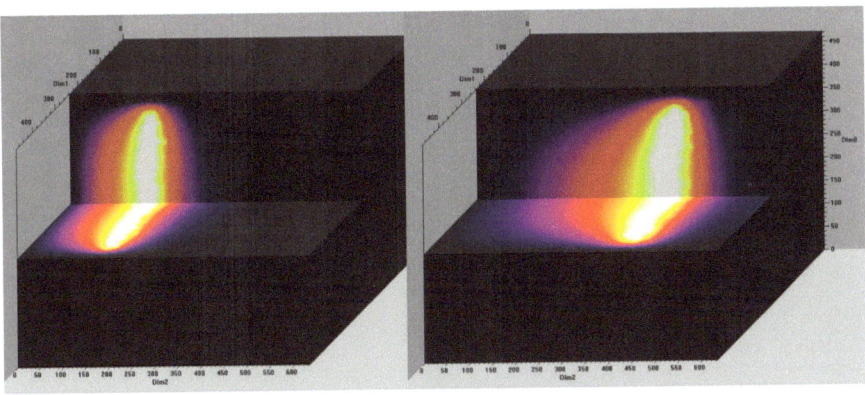

Figure 84. As described in the text. Minimal gas flow from left to right.

In Figure 84, the simulated elapsed time of the run from the left panel to the right is 3.4 hours and 7.8 hours, respectively. The image in the left panel of Figure 84 is like the portrayal often given for canister reaction zones (compare to Figure 122.) The image on the right side of Figure 84 is a little more realistic, showing a thermal remnant behind the most intense reaction zone (see a similar graphic in Figure 120.)

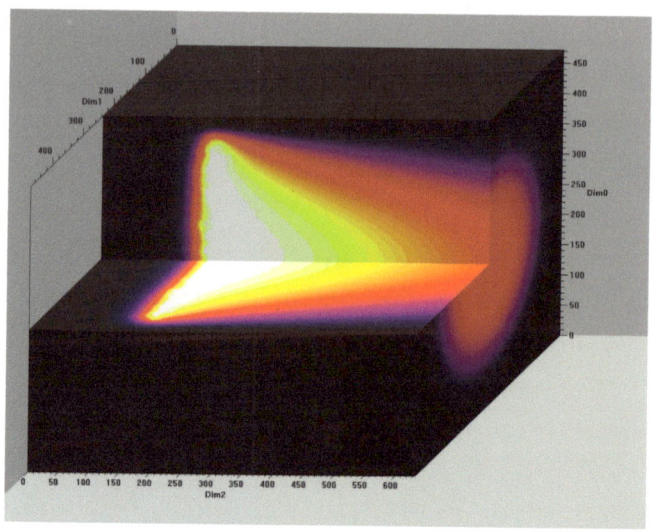

Figure 85. Adding the effect of convection from the gas flow. Elapsed time = 3.4 hrs.

Figure 85 depicts a left-to-right gas flow. Heat flows downstream, pushing the central part of the reaction zone towards the canister exit. The previous upstream "remnant" disappears.

Even though the initial conditions were deliberately unrealistic (high $\dot{V}_{CO_2}$ but low $\dot{V}_E$), there are a couple of lessons to be learned. For one, thermal sensor methods of determining canister "burn" rate tend to assume a reaction profile like Figure 84. However, as Figure 85 and other illustrations in this monograph show, ventilation rate and heat transfer to the periphery of the canister can significantly distort thermal profiles.

Potentially, distortion of thermal profiles during high CO_2 input might fool TempStik or other thermal sensors. High-temperature gases downstream of the active reaction front could lead to a premature indication of canister breakthrough. However, this did not seem to be a problem in realistic studies of thermal indicators conducted by Silvanius et al.[32].

It is well known that for sodalime to be effective at absorbing CO_2, the moisture content must be constrained to narrow limits (~14-19%). Excess heat can dry out absorbent, which could account for the drop in canister duration at the higher temperatures shown in canister duration tests at NEDU (Figure 26 in Chapter 4).

Cellular Automata

With some of the many possibilities of the SPM model being revealed to this point, it is worthwhile to remind the reader that the SPM is based on 288,000 cellular automatons. Automatons are conceptual cells whose physical and chemical state at any moment in time is affected by the average states of its neighbors.

In this case, each cell is influenced by its six adjacent neighbors (six since this is a simulated 3-dimensional volume.) By *state*, we mean many, such as temperature, presence or absence of CO_2, number of absorption sites filled or empty within each cell, quantity of calcium carbonate in each cell, and etcetera.

Since this author is neither a computer scientist nor mathematician, I was blissfully ignorant of the interesting history of automatons at the time of the SPM's first publication in 2001. As already mentioned, Wolfram[24] published his work a year later, but due to my interest in physical science and physiology, I still missed the historical connection, until recently.

I also missed all the controversy over automatons, primarily regarding their applicability to wide-ranging fields of science and mathematics. That is just as well. I spent my time investigating what the SPM taught me, uninfluenced by other's opinions.

What is important is that I find this application to be well suited to the problem at hand, studying the thermokinetics of scrubber canisters. As will be seen in the next chapter, the SPM seems to do its job exceptionally well.

The SPM is an example of a class of automata called a stochastic cellular automaton, or probabilistic cellular automata (PCA). That is why I sometimes refer to it as a Stochastic Physical Model.

Probability enters the system in several ways. The first way is through uncertainty as to which cell at the scrubber entrance will receive a CO_2 molecule. As many as 3600 of them receive one molecule simultaneously at the maximum CO_2 injection rate. For any lesser CO_2 production, however, the probability of an entry cell gaining a molecule is dependent on both the CO_2 load and a number chosen by a pseudo-random number generator built into Microsoft Visual Basic 6.0.

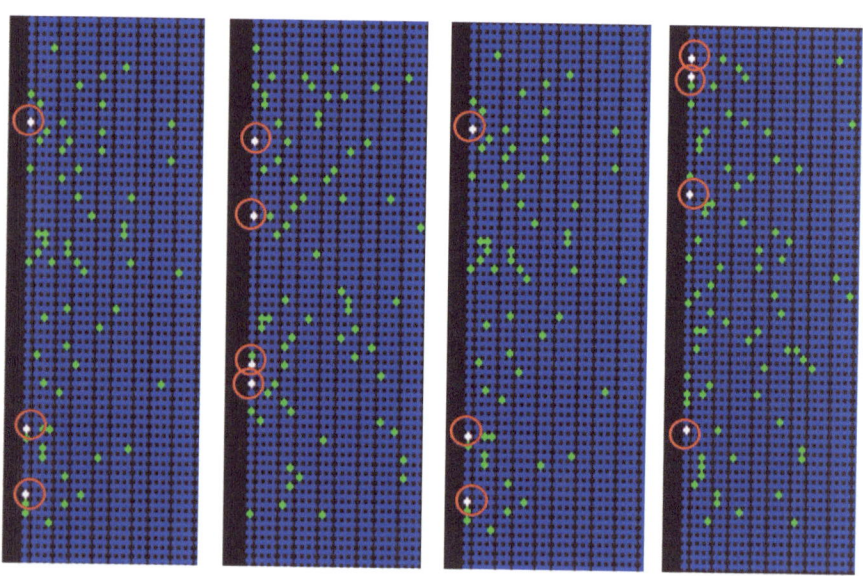

Figure 86. Red circles surround white dots representing new CO_2 molecules entering a scrubber canister.

Figure 86 shows four frames from the early stages of four different simulation runs in 40°F absorbent beds in 40°F water. The white dots circled by red rings are new CO_2 molecules entering the canister at random locations in one of 60 slices through the canister. CO_2 absorption sites are indicated by green. Luck determines whether the flowing CO_2 will be absorbed at some point or pass through the canister to be inhaled.

The second effect of probability in the stochastic model comes from the odds of the CO_2 *molecule* finding a free absorption site within each cell. The identifier for each potential site is again chosen by a random number generator. Finally, the probability of each molecule being absorbed is determined by the cell's temperature and the corresponding reaction probability as defined by the Arrhenius-like curve in Figure 42.

As a reminder, all the above must happen within the allotted residence time, a function of the average flow rate. If a vacant absorption site is not found within time constraints, then the simulated molecule moves forward to the next cell downstream.

Summary

To reiterate the statement of synchronicity at the beginning of this chapter, this computer programming effort began with a detailed description of a three-dimensional space filled with cuboidal, non-biological cells. Most cells contained chemically reactive material, with each cell exposed to a stream of chemical reactants, i.e., CO_2. Some of the cells were chemically passive, containing water to absorb heat from the non-water cells.

The "real-time" 2-dimensional displays like those in Figure 86 were intriguing, but the full power of the simulation was released by 3-D imagery. That required the post-processing tool Slicer Dicer (Visualogic.)

I then realized I had stumbled upon something interesting and worth sharing. At the time, I did not know that an entire field of mathematics existed devoted to what is properly called cellular automata. To be honest, if I had known, I would have thought it irrelevant. I was describing simple physics, which resulted in complex results given enough computational space to operate. To me, the approaches seemed totally different.

Going forward, I realize that such a parochial approach is not ideal. Having first grounded themselves in the cellular automata works of Stephen Wolfram and others, a dedicated student may be able to extract even more insight from the SPM. For that reason, I highly recommend the book.

CHAPTER 6. APPLICATIONS

Cold-Soaked Canisters

Navy operational data indicated that cold-soaked rebreather canister reactions were not always self-sustaining. The canister would begin absorbing CO_2 but soon quit, exposing the diver to high CO_2. As always, NEDU investigated.

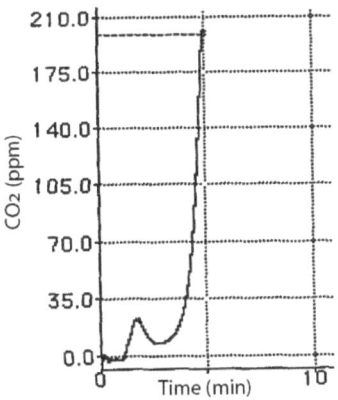

Figure 87. Laboratory test of CO_2 absorption in a cold canister.

The black curved line in Figure 87 is a plot of CO_2 exiting a cold-soaked NATO test cylinder with the CO_2 concentration in parts per million. (200 ppm is 0.02%.)

NATO specifies a particular test fixture[18] to be used for routine absorbent quality testing. That fixture is a calibrated glass tube, an "Activity Tube" as specified in STANAG 1411. It is 30 mm in diameter and contains 105 mL of test absorbent. Surrounding that test fixture, NEDU added a water bath, maintaining the outside of the cylinder at a constant cold temperature (35°F).

Typically, when filled with good quality absorbent, it usually takes about 40 minutes for the NATO Activity tube to "breakthrough." However, the average breakthrough occurred in only seven minutes when the fixture was wrapped in a water jacket and cooled to 34°-36°F for this test.

"Exhaled" CO_2 broke through the test cylinder almost immediately, then partially recovered briefly before passing straight through the test bed by the time 5 minutes had elapsed. The initial heating of the granules caused some activation of the absorption reaction, but the exothermic absorption reaction could not be sustained. The CO_2 exiting the fixture rapidly rose and never returned to low levels.

Figure 88 is the SPM simulation result which shows an immediate failure of the canister at 35°F, followed by partial recovery, and then a complete failure. The reaction front could not be sustained except at warmer temperatures.

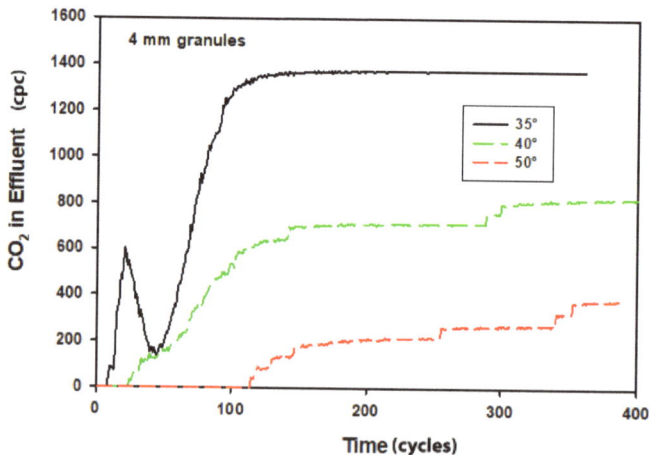

Figure 88. Transient scrubber recovery before complete failure.

The "cpc" units on the vertical axis for Figures 88 and 89 are specific to the SPM model, and stand for overflow counts per computational cycle. A short warmup breathing period on a cold canister may not guarantee sustained scrubber performance.

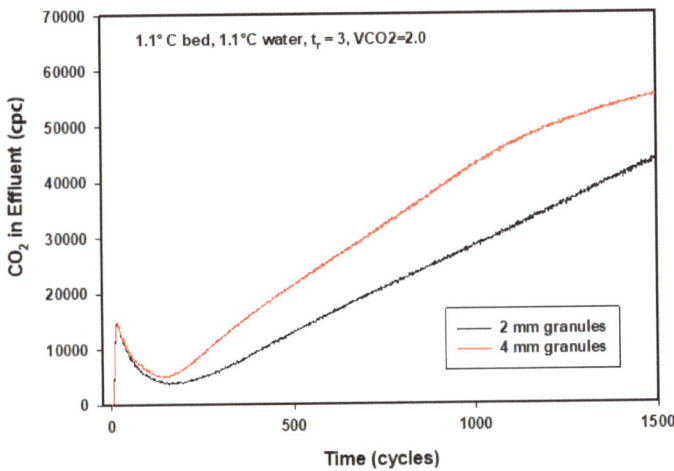

Figure 89. The simulated effect of granule size on premature breakthrough in pre-chilled canisters.

Interestingly, the SPM predicts that a bed of large absorbent granules fails more robustly than a bed of smaller granules (Figure 89).

The following frames from the simulation (Figure 90) show the effect of cold soaking a canister. Unlike the first sequence of figures (Figures 43-47), the interior of the canister is at a uniformly cold temperature. Although the reaction begins, the heat of reaction soon begins to wane, with the hottest part of the reaction contracting in size. As heat is pulled away by the chilled granules, robbing the reaction of energy, more and more CO_2 passes through the canister without being absorbed. In the common parlance, CO_2 is breaking through the canister.

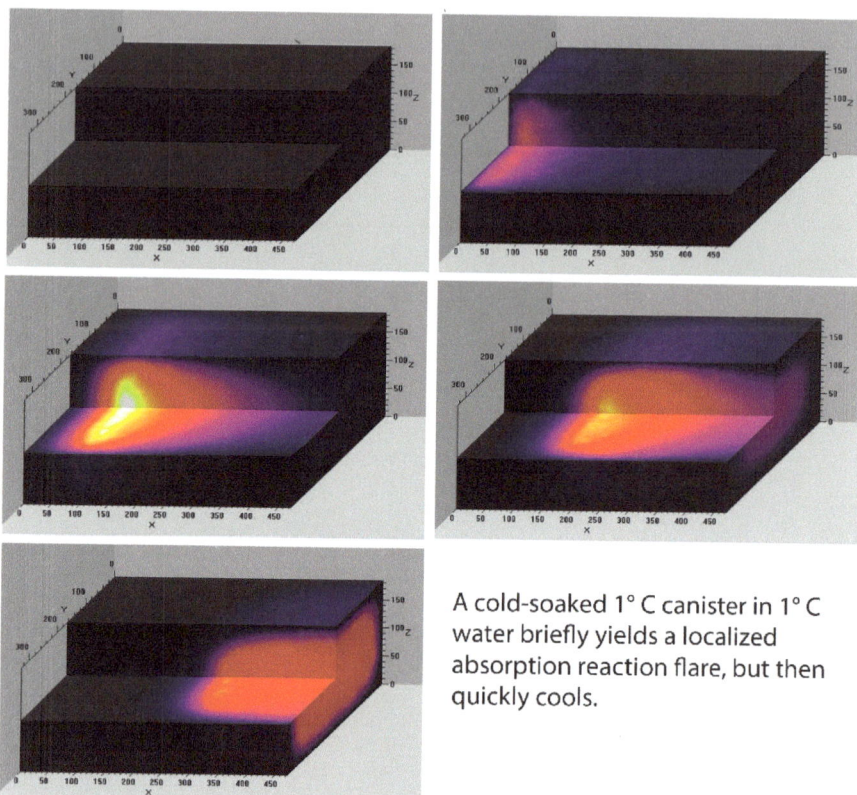

Figure 90. A cold-soaked canister was not able to sustain its reaction intensity.

Thermal Conductivity

Unlike the result of the previous simulation, when thermal conductivity and the heat transfer coefficient were reduced to 10% of their previous value, the reaction front, once initiated, was sustained (Figure 91). The heat of reaction remained close to the reaction front rather than being conducted away.

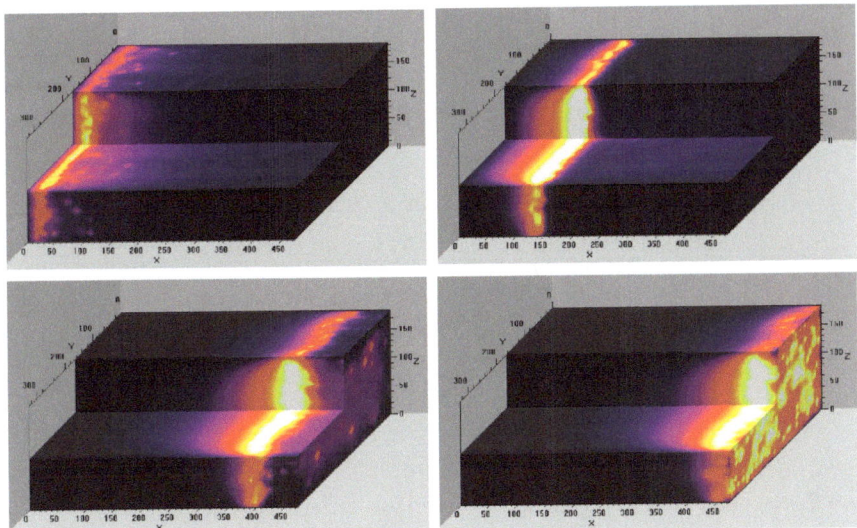

Figure 91. SPM sequence for the pre-chilled canister with lower granule conductivity.

This result shows that in the previous instance with a pre-chilled canister, in qualitative accordance with Fourier's law of conduction,

$$\dot{Q} = -k \cdot A \cdot \frac{dT}{dx}$$

the combination of large thermal gradients ($\frac{dT}{dx}$) and high thermal conductivity (k) causes the rate of heat transfer ($\dot{Q}$) to raise enough to remove heat from the reaction front faster than exothermic reactions can generate. The net result is reaction front cooling and diminished reactivity.

As always, convective heat loss also contributes to cooling and diminished reactivity.

Absorbent Granule Size Distributions

Before NATO military testing laboratories perform a canister duration test on a rebreather, samples from the absorbent pails must pass a rigorous set of qualification tests, as defined in NEDU Technical Manual 02-01, STANAG 1411 and its replacement, ADivP-03[33]. Those tests include checks on the absorbent moisture level, granule size distribution, and the amount of dusting (friability).

Some of the more egregious failures in those absorbent quality tests have occurred in the distribution of sodalime granule sizes. As demonstrated in this text, that can degrade breathing resistance and canister duration.

NATO STANAG 1411[5] calls out the following requirements for granule distributions.

Table 8. NATO allowable particle distribution.

U.S. Sieve Size	Requirement	
	4-8 Mesh	10-14 Mesh
<2	0% max	--
2 – 4	7% max	--
4 – 8	77% max	--
8 – 30	15% max	--
<7	--	1% max
7 - 10	--	30% max
10 – 14	--	48% max
14 - 30	--	20% max
>30 (dust)	1% max	1% max

To test whether sodalime meets the NATO specifications, absorbent is placed in the top of a stack of sieves. A shaker/tapper machine shakes the sieves and strikes the top of the stack of sieves, allowing granules to fall into the stack, sorting themselves by size as they penetrate down to ever-smaller sieve mesh sizes.

Table 9 translates sieve sizes into ISO measurements in millimeters.

Table 9. Mesh size comparisons.

U.S. Sieve (ASTM E-11)	Tyler Sieve	British (BS 410)	Aperture (mm) ISO std.
3.5	3.5	3	5.60
4	4	4	4.75
5	5	--	4.0
6	6	5	3.35
7	7	6	2.80
8	8	7	2.38
10	--	8	2.00
12	10	10	1.70
14	12	12	1.40
30	28	25	0.60
40	35	36	0.425

The bottom pan is a dust collection pan, where material smaller than the smallest mesh accumulates.

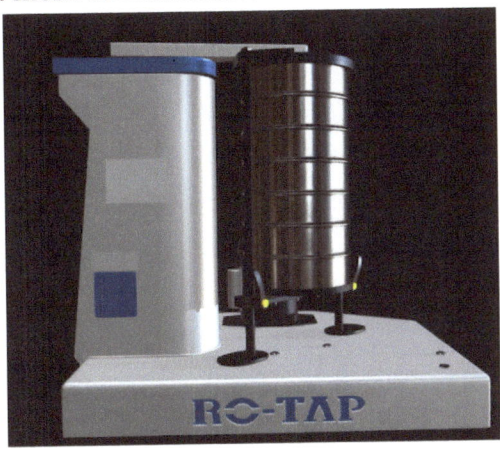

Figure 92. Ro-Tap Sieve Shaker.

NEDU TM 02-01[18] calls for the following screens: U.S. 3.5 mesh, 5, 8, 10, 14, 30, and pan—no mesh.

Out of a 100-gram sample, the percentage retained in each of those screens is recorded. One example of how that recording and analysis can occur is through NEDU proprietary software, *MeshFit*.

Below are screenshots from *MeshFit* sodalime analysis software.

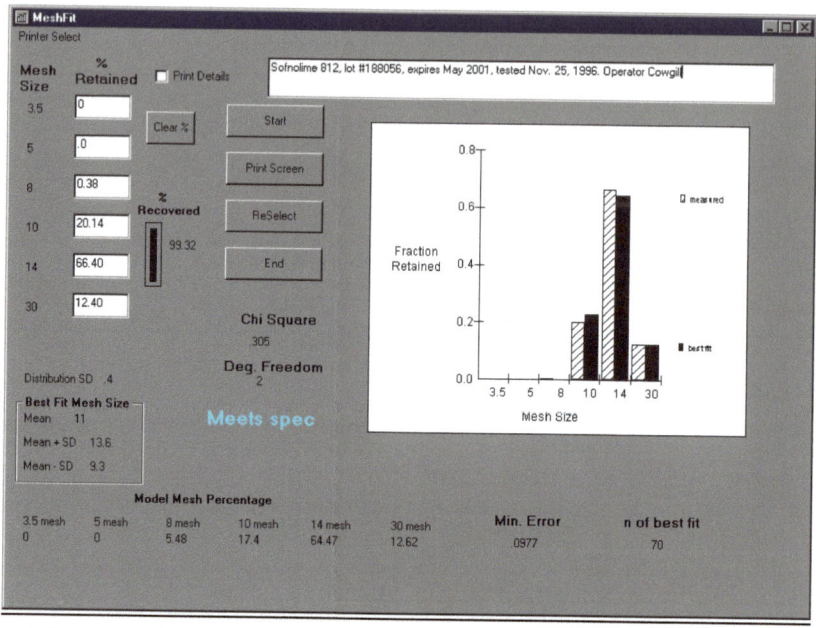

Figure 93. MeshFit analysis of a fine grain Sofnolime sample.

The NATO specifications are relatively unspecific; many size distributions can meet the NATO requirements. In Figure 93, the black vertical bars represent the makeup of the one distribution closest to the measured sample while still meeting the specification. The hashed bars represent the results of the sieving of the test sample. The black and hashed distributions' degree of match or mismatch is calculated through the Chi-Square test. The Chi-Square statistic determines whether the 100-gram sodalime sample matches the NATO specification.

In Figure 94, there is a statistically significant mismatch between the test sample and the closest match to the specification. The Chi-Square was 34, ten times larger than in the previous example, Figure

93. The conclusion was that the sample and the specification did not match.

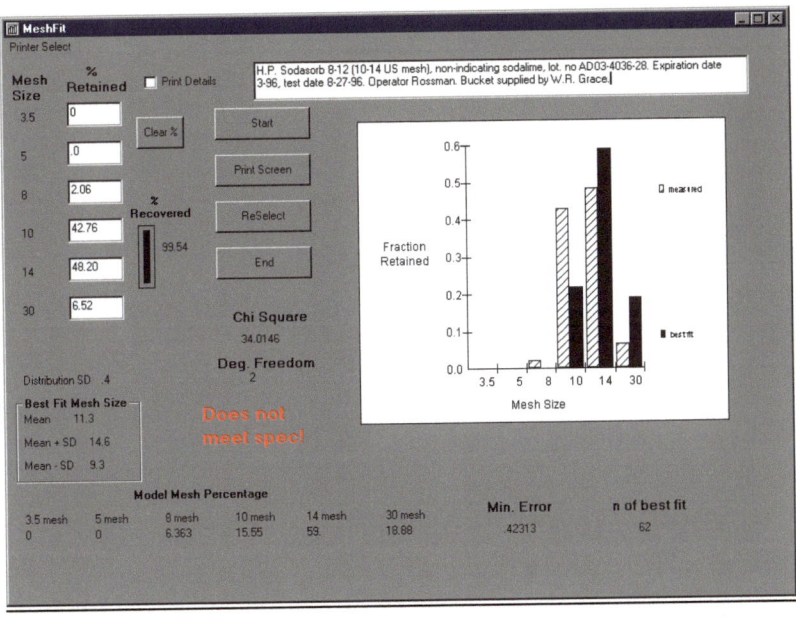

Figure 94. MeshFit analysis of a so-called fine-grain H.P Sodasorb sample that failed NATO specifications.

This discussion reinforces that you don't always get what you expect[w] regarding CO_2 absorption.

Another case in point, Harvey, Pollack, et al.[34] performed canister duration tests in an Inspiration Rebreather (AP Diving, Helston, Cornwall, UK), comparing Sofnolime 797 (Molecular Products) to Spherasorb (Intersurgical, Berkshire, UK.) Spherasorb,

---

[w] *The consequences of a government agency rejecting a manufacturer's product is fraught with potential legal and good-will consequences. Since NATO specifications are somewhat ambiguous, MeshFit was designed to provide a clear and statistically defensible answer to the question of whether the product is "fit for purpose."*

designed for anesthesia machines, has reportedly been used by some rebreather divers.

Those authors found that the time to the breakthrough of Spherasorb was considerably shorter than for Sofnolime. Spherasorb and Sofnolime 797 were not interchangeable, despite common belief to the contrary.

Friability

Absorbent hardness is an indication of abrasion resistance of the sodalime granules. While hardness figures are sometimes published by manufacturers as a percentage (e.g., 80-95%), the NEDU measures friability, not hardness. Friability measures the ease with which granules can be broken down into smaller particles and dust.

The Schlegel friability test developed at NEDU is performed as follows. A 100-gram sample of granular absorbent is placed in the uppermost sieve of the Ro-Tap shaker and sieved for 5 minutes. The sieves and pan are weighed as before, but the sieves are returned to the shaker without emptying their contents. The Ro-Tap shaker is then run for 55 minutes before the sieves are removed and weighed.

Figure 95 shows a test result for one bucket of 4-8 mesh sodalime. The pooled results from two samples are displayed on a logarithmic scale to accentuate the smaller granule sizes.

After a total of sixty minutes of shaking, some of the largest granules had broken down, adding to the 8 mesh granules and the amount of dust in the pan. Although a small number of granules were captured in the smallest sieve after 5 minutes, a large portion of that approximately 1 gram of material fell into the dust pan after the completion of the friability test.

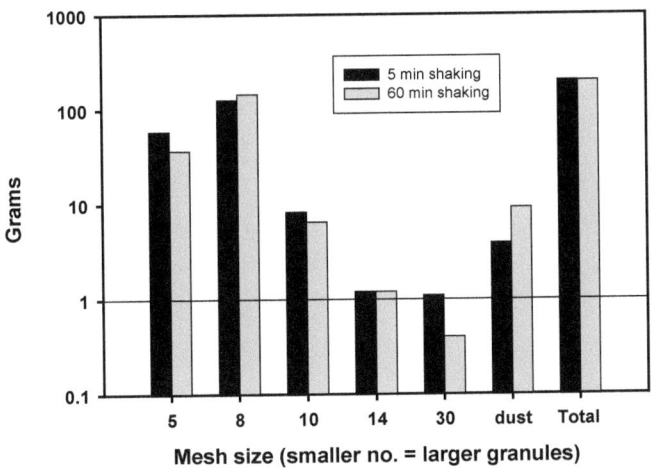

Figure 95. Friability test results of a Sodasorb sample.

After this test, this particular absorbent sample was considered "soft."

Absorbent Lot Variability

So far, the topic has been variability in performance between sodalime manufacturers. The following incident resulted from inconsistency within a single product and negatively affected fleet operations.

In response to a Fleet query about shortened canister durations in the cold, NEDU analyzed variations in absorption duration among lots of a single product in keeping with the guidance of NATO STANAG 1411[5].

Fifty-four NATO *activity* runs were performed in triplicate on 18 different lots of large grain sodalime, Sofnolime 408[35]. As described above, three of the 18 lots did not meet the British specification for Sofnolime 408. More importantly, test duration was negatively correlated with the percentage of absorbent captured on a #5 U.S. sieve when each lot was sieved with NATO-approved nested sieves. It was positively correlated with the percentage of material captured on a smaller mesh, #8 U.S. sieve[10].

Granules larger than 4.0 mm are captured on a #5 sieve, and those larger than 2.36 mm are captured on a #8 sieve. So, at cold temperatures, at least in the NATO test fixture, CO_2 absorption duration was clearly favored by smaller granules. Details of this study can be found in reference (35).

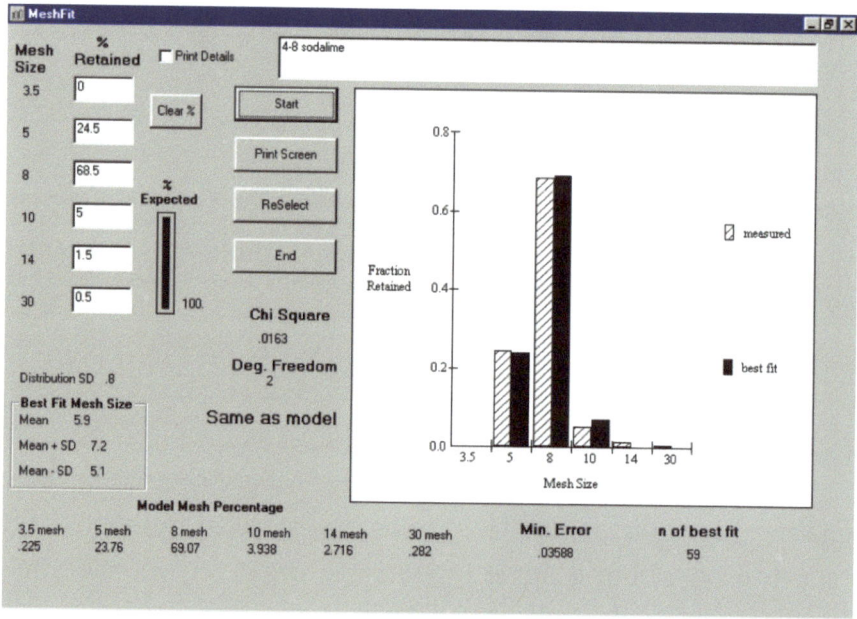

Figure 96. The measured and expected distribution of Sofnolime 408 granule sizes match.

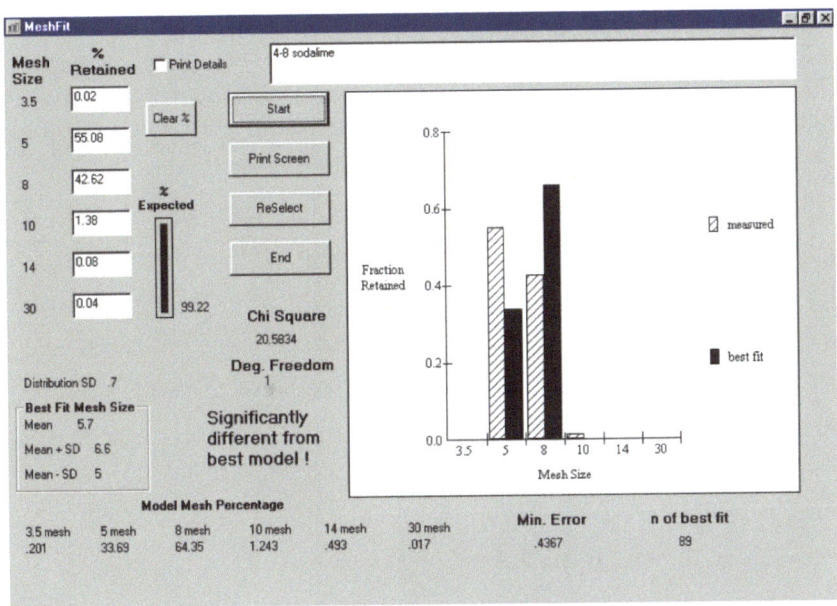

Figure 97. The measured and expected distribution of Sofnolime 408 granule sizes <u>did not</u> match. Granule sizes were too large.

This sodalime analysis was conducted two decades ago. Lot variability <u>may</u> no longer be an issue. However, there is value in spreading a small sample of absorbent in your hand before trusting your life to it. If something looks different, or smells different, be suspicious. Absorbent problems have cropped up at least three times in the past. It could happen again.

Specialized Canisters

In 2001, two NEDU investigators explored a new scrubber canister design for the Submarine Rescue and Decompression System (SRDS). The existing scrubber had a large rectangular entry boundary for CO_2 but an offset, circular exit on the other end of the canister[36].

In Figure 98, the computed canister temperatures within the investigational absorbent bed were <u>projected</u> onto the three planes of the simulated rectangular canister. As usual, CO_2-containing gas flowed from the left of the illustration to the right.

Although CO_2 entered the canister equally distributed, the asymmetrical gas flow path swept the heat from the planar reaction path into a comet-like tail headed toward the off-center exit (upper right of the horizontal plane). Such asymmetrical heat distribution would make a linearly arranged temperature sensor like a TempStik nearly useless for predicting impending canister depletion.

This work evolved into a Warkander patent[x] for his thermal scrubber monitoring system initially used by license in the rEvo, Sentinel, and Hollis Explorer rebreathers.

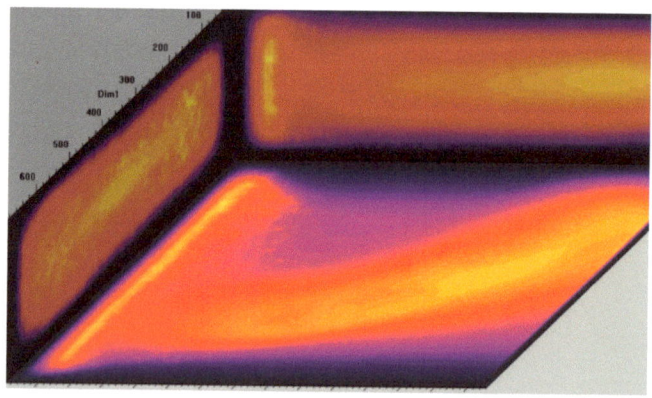

Figure 98. The effect of asymmetrical convection on thermal distributions.

---

[x] *U.S. 6,618,687. Temperature-based Estimation of Remaining Absorptive Capacity of a Gas Absorber, September 2003.*

Predive Decision Making

Figure 99. Leon Scamahorn and Becky Kagan Schott with their ISI Megalodon Rebreathers.

Suppose you plan a dive to 1000 feet seawater (306 msw) to identify a shipwreck rumored to contain something of interest. Due to the great depth, you want nothing but the best for your planned dive. As far as sodalime goes, you believe fine-grain sodalime is the best "sorb" for your mission. It's the most expensive on the market but completely worth it, you think, because of the greater CO_2 absorption capacity it has compared to large grain sodalime. That is especially true in cold water, so you hear. Because of that added absorption capacity, your canister scrubber should last longer than if using any other absorbent.

However, after reviewing the results of other deep dives, you are second-guessing your first decision. Fine-grain sodalime is harder to breathe through than large grain absorbent. However, you only plan to be on the bottom for five minutes, just enough time to take a few photos.

Will you be making a sound decision using the "best" sodalime? Well, in the following figures, we will examine the consequences of choosing fine-grain sodalime versus larger grain sodalime.

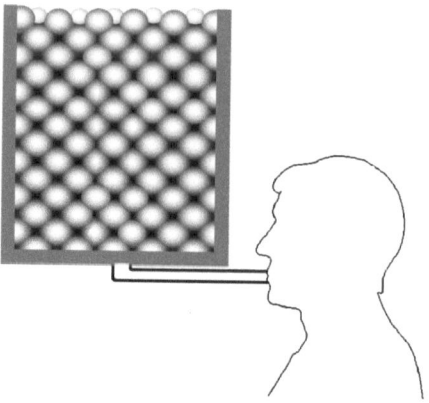

Figure 100. Illustration of breathing through pores in a scrubber canister.

Figure 100 is a conceptual model of the problems encountered by a diver breathing through a bed of spherical particles of varying sizes. The diver's breath travels through the void spaces between sodalime granules, which in this case are large because of the large spheres illustrated. As the spheres or granules become smaller, the pressure drop across the canister increases markedly, especially at higher flow rates, as shown in Figure 100. But how much more does it increase?

M. Leva[37] is only one of many civil engineers who have quantified the pressure drops across packed beds of uniformly sized spherical particles. The predictions from his equation are plotted in Figure 101. A diver is hoping for a small pressure drop across the canister, not the opposite.

134

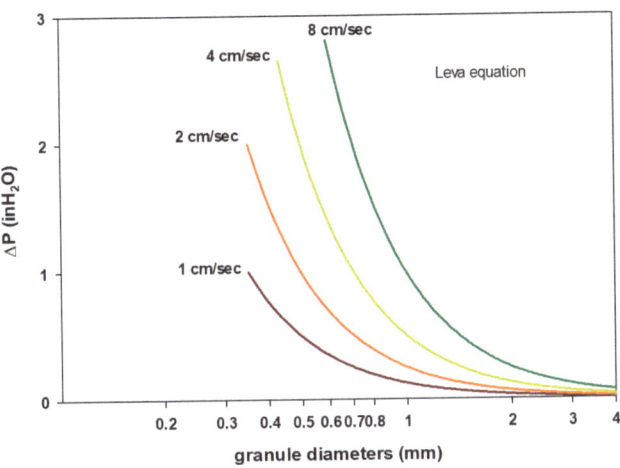

Figure 101. Predicted pressure drop across an absorbent bed based on Leva's equations.

In keeping with civil engineering convention of the time, pressure drop was measured in units of inches of water (inH$_2$O.) Physiologists tend to use centimeters of water (cmH$_2$O.)

Leva's work was not particular to rebreather scrubber canisters. However, U.S. Navy engineers produced remarkably similar graphs when studying rebreather canisters[37]. Therefore, these types of graphs are broadly applicable to our discussion.

As we have seen multiple times in this monograph, sodalime granules are not identical in size. While the average granule size may be small (or large), there is a wide assortment of granule sizes in any one sample of sodalime. And that assortment means smaller granules can fill the gaps between larger granules, leaving the diver with a less porous canister than intended. With fewer and smaller pores, breathing resistance rises remarkably.

In Figure 102, modified from Bear[39], the Greek letter ϕ represents the porosity of beds containing well sorted or poorly sorted spherical particles. A canister with a porosity of 32% would be far easier to breathe through than one of 17% or 12.5% porosity.

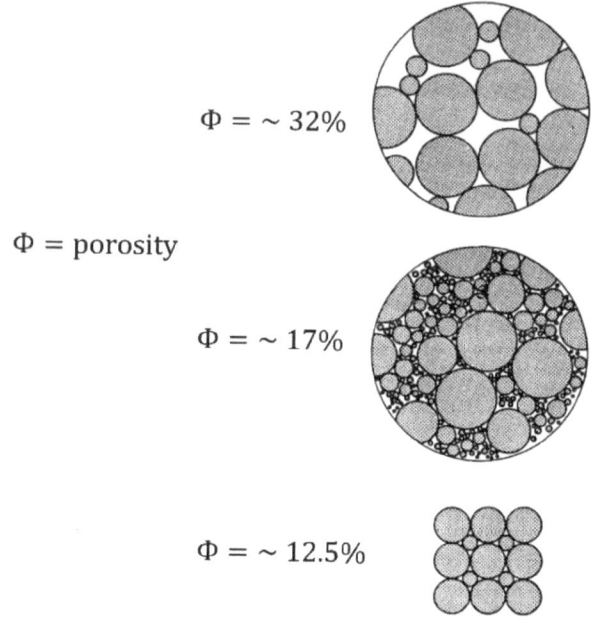

Figure 102. Beds of uniformly sized granules have higher porosity (ϕ) than beds with a distribution of granule sizes.[39]

Ideally, you would want the most absorption activity possible from your scrubber. And as we've already mentioned, that comes from using small granules. However, small granules increase breathing resistance. And confusing the issue is the question of how widely distributed the granule sizes in your canister are when the dive begins.

Figure 103 is a notional histogram of normally distributed sodalime granules. Recalling the granule size analysis presented at the beginning of this monograph, our single Sofnolime 812 sample met such a normal (or Gaussian) distribution. The mean granule size for the illustrated sample is 1.75 mm. The distribution of sizes is symmetrical, with a standard deviation (σ) of 0.4 mm.

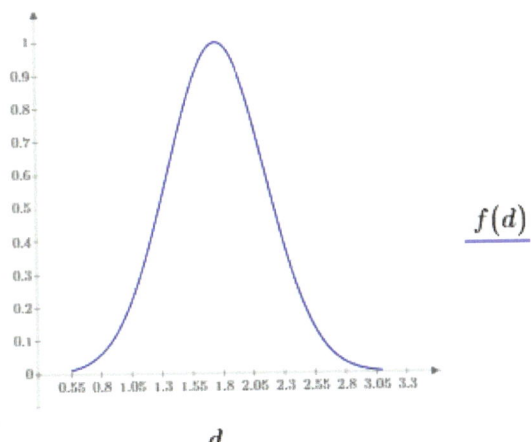

Figure 103. A Normal distribution of granule sizes for a mean diameter of 1.75 mm and a standard deviation of 0.4 mm, is appropriate for Sofnolime 812 granules.

For the following graphs, the equations for estimated absorbent activity and flow resistance are solved over a range of both mean granule diameters and standard deviation of the mean. In other words, those equations are treated to all possible combinations of absorbent bed conditions over the range of interest.

$$Activity_{est} = \sum A_{sg} \cdot \frac{1-\phi}{\sum V_{granule}} \qquad (12)$$

The math is straightforward when assuming spherical granules. The surface area available for absorbing CO_2 is easily computed. Likewise, the volume of a spherical granule is known. In this analysis, the estimated *Activity* is nothing more than the ratio of total granule surface area for a given total granule volume. Porosity determines how tightly the absorbent bed is packed.

Based on the work of Dexter and Tanner,[40] a linear relationship between ϕ and σ was assumed in the expression for granule packing density ρ_p. Specifically,

$$\rho_p = 0.52 + 0.8 \cdot \sigma^1$$

and

$$\phi = 1 - \rho_p$$

Consequently, porosity ranged from 0.48 when σ was 0.0, to 0.16 when σ equaled 0.4.

Figure 104 shows how estimated absorption activity varies between the fine-grain absorbents shown in Figure 8 and the large grain sodalime. The smaller grain absorbent has more activity, as defined in Equation (12). Interestingly, the estimated difference is accentuated at small standard deviations.

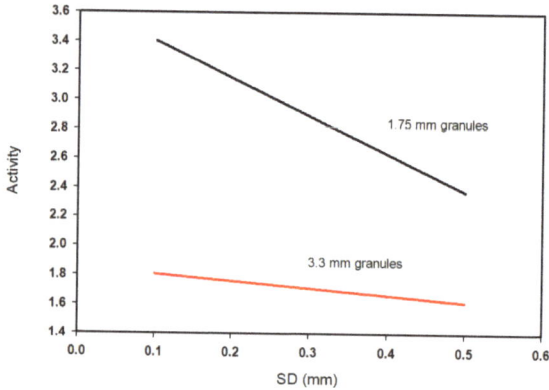

Figure 104. Activity vs σ (S.D.).

Resistance Matters

As a child, I made a useful discovery. One straw was all I needed to sip lemonade. But it was easier to drink a milkshake using two straws rather than just one.

Figure 105. Boy drinking a milkshake. (Stock photo.)

Milkshakes were thicker than lemonade, and the effort required to suck up a mouthful of chocolate malt was considerably eased by doubling up the straws. I didn't know the correct terminology then, but I had intuitively learned how to reduce the resistance to flow of a viscous liquid.

Not everything has to be explained by math. However, a little engineering math is required to appreciate the numbers you're about to see. This math is brought to you by French physicist and physiologist Jean Poiseuille and independently by Gotthilf Hagen. To give full credit, it's called the Hagen-Poiseuille equation, and applies to incompressible fluids in cylindrical pipes of a constant cross-section. It does not apply to scrubber canisters, per se, (we'll get to those more complex equations shortly), but Poiseuille's Law is used here to define flow resistance in simple terms.

ΔP, pressure drop along the length (L) of a tube is,

$$\Delta P = \frac{8 \cdot \mu \cdot L \cdot Q}{\pi \cdot r^4} = \frac{8 \cdot \pi \cdot \mu \cdot L \cdot Q}{A^2}$$

where μ is dynamic viscosity, Q is volumetric flow (liters/s), and r is tube radius. A is the cross-sectional area of the tube.

After rearrangement, flow resistance (R) is equal to,

$$R = \frac{\Delta P}{Q} = \frac{8 \cdot \pi \cdot \mu \cdot L}{A^2}.$$

In the above straw example, flow resistance increases the longer the straw, the more viscous the fluid[y] (milkshake vs. lemonade), and the smaller the tube's cross-sectional area. The higher the resistance to flow, the more negative the required mouth pressure must be to produce a fixed volumetric flow rate. As a child, I reduced resistance by increasing the combined cross-sectional area of the straws.

R has units of pressure over volumetric flow. For physiological purposes, units of $kPa \cdot L^{-1} \cdot s$ or $cmH_2O \cdot L^{-1} \cdot s$ are preferred. (A kilopascal (kPa) is about 10 times larger than a cmH_2O.) The estimated resistance of pulling air (μ=0.0813 cps) through a straw is 0.52 $kPa \cdot L^{-1} \cdot s$ and 5.3 $cmH_2O \cdot L^{-1} \cdot s$. For water, μ=1 cps and R=6.4 $kPa \cdot L^{-1} \cdot s$. For a so-called "optimal" milkshake[41] (μ=725 cps), R=4.62 · 10^3 $kPa \cdot L^{-1} \cdot s$.

Scrubber Flow Resistance

To calculate the flow resistance in a rebreather scrubber, the Ergun equation[42] can be used (Appendix B.) L is the length in cm of an assumed cylindrical scrubber canister, D is the cylindrical radius

---

[y] The viscosity of water is 1 centipoise (cps.) The optimal viscosity of a milk shake is 725 times higher[41]. R, flow resistance, is also 725 times higher. Yet somehow, it's entirely worth it.

in cm, and ϕ is absorbent bed porosity[z] as affected by the standard deviation of granule sizes (in mm). As before, μ is gas viscosity, and ρ is the gas density of a 1.3 PO$_2$ heliox gas mixture at 306 msw, 1000 fsw (6.0 gm/L at 37°C.)

We solve for the case of 75 lpm diver ventilation at 1000 fsw. For a simulated bed of Sofnolime 408 (μ=3.3 mm, σ = 0.2 mm), the calculated pressure drop across the canister was a tolerable 2.1 cmH$_2$O or 0.2 kPa.

Flow resistance was 0.05 kPa·L^{-1}·s which is low considering the calculated resistance of breathing through a standard straw with the same helium viscosity of 0.019 cps is about 0.12 kPa·L^{-1}·s, according to the Hagen-Poiseuille equation.

For sodalime absorbent beds with granules the size of Sofnolime 812 (mean diameter, μ=1.75 mm, σ = 0.4 mm), the calculated pressure drop across the canister was a staggering 58 cmH$_2$O or 5.6 kPa. During breathing, the predicted peak-to-peak pressure swing, ΔP, was 115 cmH$_2$O. Calculated flow resistance was 1.4 kPa·L^{-1}·s, or 14.7 cmH$_2$O·L^{-1}·s.

The best way to put that amount of resistance into perspective is to once again employ the Hagen-Poiseuille equation[aa]. Assume you are breathing through a modified straw. It has the usual diameter of 6.1 mm (~ ¼ inch.) But it would have to be 2.56 m (8.4 ft) long to possess the same breathing resistance as that in a canister of well-packed 8-12 grade sodalime absorbent.

---

[z] *Whereas we use ϕ as the symbol for bed porosity, Appendix B uses the symbol ε as cited in the quoted reference.*

[aa] *The Hagen-Poiseuille equation applies to incompressible fluid. However, at very low physiological pressures, the compressibility of air is negligible.*

Resistance Limits

Silverman[43] specified that breathing apparatus used at high ventilatory rates should have an inspiratory resistance no greater than 4.5 cmH$_2$O·L^{-1}·s (0.44 kPa·L^{-1}·s) and an expiratory resistance no higher than 2.9 cmH$_2$O·L^{-1}·s at a flow rate of 1.42 L·s^{-1}. The units given by Silverman are true units of resistance: pressure divided by volumetric flow.

Ten years after Silverman's 1945 report, Harvard's Jere Mead[44] addressed breathing resistance at increased ambient pressures, stating that total breathing resistance, meaning internal (that within the diver's airways) plus external (that due to a UBA), should be no greater than 12 cmH$_2$O·L^{-1}·s (1.18 kPa·L^{-1}·s). He suggested that during air breathing, internal resistance alone would reach this value at 265 fsw. Both Silverman and Mead agreed that increases in internal resistance due to increased flow rate or high gas density led to reductions in tolerated external resistance.

Figure 106. Dr. Jere Mead, Harvard

Because of its simplicity and appropriateness to the subject, Mead's proposed limit of 12 cmH$_2$O·L^{-1}·s is used in the following graphs. *(It is not the direct source for the Navy's UBA acceptance limits. However, the concept espoused by Mead* **is** *the source for the Navy's limits.)*

To see how Mead's proffered 12 cmH$_2$O·L^{-1}·s limit applies to limits for the external resistance of UBA, we must know how internal (pulmonary) resistance varies over a range of depths and gas densities. Thanks to a U.S Navy saturation dive at NEDU to 1500 fsw (~450 msw), I was able to obtain that data in six heliox-breathing divers[45].

Shortly after that study of Navy divers, I sought to validate Silverman's and Mead's hypothesis that increases in gas density and internal resistance lead to decreased tolerance to the external resistance provided by breathing apparatus.

Figure 107 is a summary of that work[45]. Each of the dots below represents a dive with moderately heavy exercise at a calculated gas density. ΔP is measured peak-to-peak mouth pressure. Numerous data points are superimposed on each other so not all the data is visible in this two-dimensional plot.

ΔP was measured during diving research at the Naval Medical Research Institute (NMRI) and NEDU[46-49]. That data was obtained down to depths of 450 msw (1500 fsw) and eventually used to define the probabilistic tolerance of divers to depth as a function of gas density. It is also used as an input to proprietary physiological event prediction software called *Predict*[46].

The diagonal solid black line shows the demarcation between uneventful dives (black-filled circles) and mostly eventful dives (red-filled circles). As ΔP rises above the zero-risk line for any gas density, there is an increase in the probability of an eventful dive.

An eventful dive was characterized by premature cessation of work due to loss of consciousness, or overwhelming dyspnea or

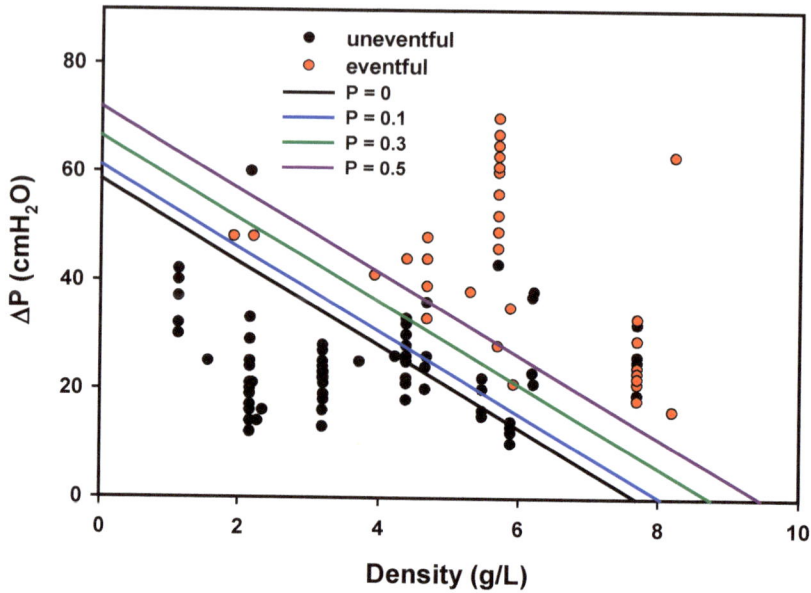

Figure 107. The result of a U.S. Navy meta-analysis of multiple Navy dives over a wide range of gas densities.

breathlessness. Isoprobability lines are drawn for untoward event probabilities ranging from zero to 0.5.

As a gas density reference, air at 140 fsw and 37°C (average lung temperature) has a density of 6 g/L, as does a helium-oxygen mixture at 1125 fsw (with a PO_2 of 0.45 ata). For the same temperature, air at 200 fsw has a density of 8 g/L.

The Navy data in Figure 107 supports an inability to perform useful work in the density range predicted by Mead, air breathing at approximately 265 fsw with a gas density of roughly 9.6 gm/L.

Returning to the subject of rebreather canisters, in the next two figures we see how calculated canister resistance affects overall tolerance to breathing resistance.

In Figure 108, the average pulmonary resistance measured in resting divers by the interrupter technique[45], is plotted as a straight blue line. Mead's proposed limit (red line) is constant with depth. His thinking (and mine) was that on average respiratory muscles can only tolerate a given amount of resistance regardless of the environment you're inhabiting.

At a ventilation rate of 40 lpm corresponding to moderate work, the flow resistance added by a canister with large grain sodalime only slightly elevates total resistance as gas density increases. What is left out of this figure is the unknown resistance of mouthpieces, hoses and breathing bag connections.

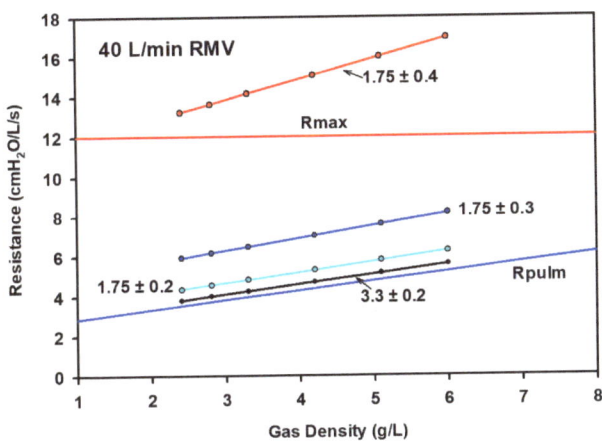

Figure 108. The calculated sum of pulmonary and canister resistance as a function of gas density, relative to Mead's proposed maximum.

Circular symbols represent calculations at gas densities of 2.4, 2.8, 3.3, 4.2, 5.1 and 6.0 g/L at a temperature calculated at 37°C. Those densities correspond to 1.3 ata oxygen in helium (heliox) gas mixtures at 200, 300, 400, 600, 800, and 1000 fsw.

At a moderately-hard ventilation rate, fine grain sodalime (1.75 mm average granule diameter, comparable to Sofnolime 812) encroaches upon the so-called respiratory reserve when granule sizes have a standard deviation (σ) of 0.3 mm (dotted blue line.) When σ equals 0.4 (dotted red line), then the estimated flow resistance for the fine grain (Sofnolime 812) exceeds the Mead maximum[bb].

Large grain absorbent (dotted black line, comparable to Sofnolime 408) barely made a difference to respiratory reserve, especially with a small standard deviation (σ) of 0.2.

At 75 lpm (Figure 109), the Mead maximum is greatly exceeded when granule size is 1.75 ± 0.4 mm and reaches the maximum with a σ of 0.3 (blue dotted line.)

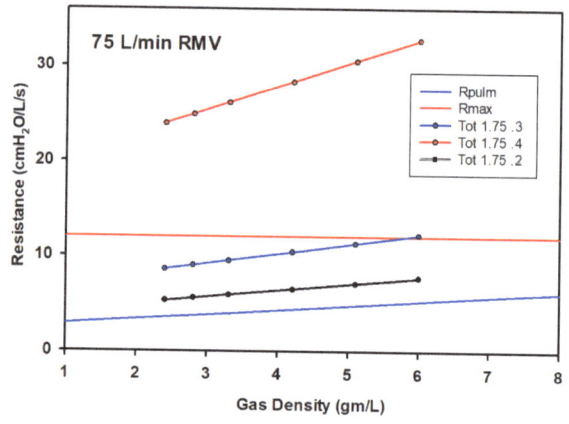

Figure 109. From left to right, resistance at 200, 300, 400, 600, 800, 1000 fsw.

Remember that these calculations assume that no matter how relaxing the dive was intended to be, a ventilation rate of 75 L/min, or more, might be required in an emergency.

To summarize, 4-8 grade absorbent would be a better choice for that ultra-deep dive. In an emergency, having a slightly longer-lasting canister does no good if you can't breathe.

[bb] *In Figure 103, the mean and standard deviation measured from the author's sampling of Sofnolime 812 was 1.75 ± 0.4 mm. Those numbers may no longer apply.*

Interactions

Physiologically, every system in your body is connected to every other bodily system. Figure 110 expresses that fact as it applies to physical exertion in stressful environments, such as deep-water diving.

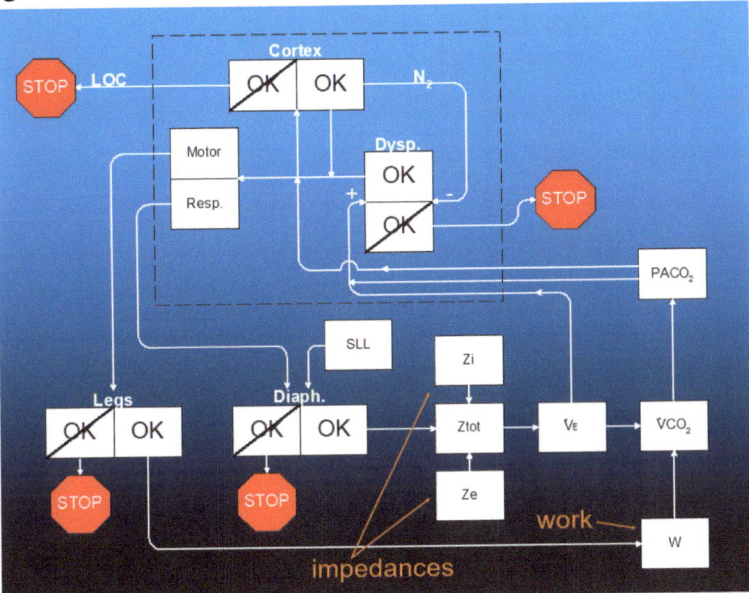

Figure 110. The physiological loop[13] and sources of untoward events in diving. The rectangular dashed line consolidates Central Nervous System (CNS) centers. SLL = static lung loading, Zi = internal respiratory impedance, Ze = external (UBA) impedance, Ztot = total impedance, W = work, Motor = motor centers in the CNS, Resp. = respiratory centers in the CNS, LOC = loss of consciousness, N_2 = nitrogen, postulated to have a salutary effect on dyspnea (Dysp.) Diaph. = diaphragm.

In this drawing, high PA_{CO2}, and ventilation ($\dot{V}_E$) were assumed to have a stimulatory (+) effect on dyspnea.

Normally, human performance operates in a closed physiological loop. There are no off-ramps and no stop signs to interrupt normal function.

But that is not always the case. Figure 110 has a lot to teach us about predive decisions and knowing where things could go wrong. As luck would have it, several things might conspire against you. The first fateful choice might be the decision to dive to 1000 fsw in a rebreather. As you have seen, even heliox becomes dense at that depth, and dense gas is a significant contributor to respiratory impedance (which includes resistance), as indicated by the boxes labeled Z and the orange arrows. Respiratory impedance directly affects ventilation ($\dot{V}_E$).

Usually, an increase in respiratory impedance results in a decrease in ventilation. But if you must unexpectedly work hard while on the bottom, then that work (W) requires an increase in CO_2 production ($\dot{V}_{CO_2}$). With an increase in $\dot{V}_{CO_2}$ and a decrease in $\dot{V}_E$, CO_2 in the blood and lungs will inevitably increase.

Two things can happen when there is the perceived need to increase ventilation but an inability to do so. One is the sensation of breathlessness, which is a sure signal to stop working if you can. (That leads to the stop sign on the right side of Figure 110.) If you cannot stop working, the growing sensation of unrelieved breathlessness may lead to panic.

Panic, as you know, leads to rapid but inefficient breathing, which may lead to a continued rise in arterial CO_2.

The second thing that can happen is that no dyspnea may occur, and loss of consciousness may occur without warning. In Figure 110, loss of consciousness (LOC) is the top left stop sign.

That was precisely what happened to the Navy diver who lost consciousness during my research, as described in the *Decoupling* section of Chapter 5. Fortunately, he was in a dry hyperbaric chamber at the time, with no risk of drowning.

Down the Rabbit Hole

I was recently asked this question of considerable import. Which is more lethal for rebreather divers, breathing resistance or CO_2? From Figure 110, you now know the answer: it's both.

Having noted my insistence that high breathing resistance is dangerous, an early reader of this manuscript asked me to convert flow resistance into something more familiar to most divers, like "Work of Breathing."

Well, that simple question does not have a simple answer. To answer it, we must first begin with definitions. What has long been called work of breathing (WOB) is more accurately termed *volume-averaged pressure* ($\overline{P_V}$). That WOB is a misnomer has been discussed in NEDU Technical Manuals 1-94 and 15-01[3,6] and the reference book chapter titled *Underwater Breathing Apparatus* in The Lung at Depth.[13]

When writing an article for the lay press, it's appropriate to write down a result and quote the source, but more is required for a technical monograph like this. Since The Lung at Depth is a technical book[13] only accessible in libraries, the math is explained as new work in Appendix C.

As for the question asked by the reviewer, calculated $\overline{P_V}$ was 8.9 kPa for the case of fine-grain sodalime, far above any allowable limits (2.41 kPa) for a depth of 300 fsw or deeper. $\overline{P_V}$ was 0.32 kPa for the large grain absorbent with a narrow dispersion of granule sizes.

The larger granules had a 28 times smaller $\overline{P_V}$. Or, using the common but inaccurate terminology, the *Work of Breathing* calculates to be 28 times less with the larger absorbent under these extreme depth and ventilation rate conditions. *(For the physicists among you, the true Work of Breathing for a ventilation of 75 L/min at 1000 fsw was calculated to be 22.2 **Joules**. See Appendix C.)*

Table 10. Resistive Effort (RE) data from rebreather testing[6].

Ventilation (L/min)	Depth (fsw)					Goals (kPa)
	165	198	231	264	300	
22.5	0.38	0.39	0.39	0.40	0.41	1.37
40.0	0.55	0.57	0.57	0.62	0.65	1.37
62.5	0.83	0.91	0.96	1.02	1.08	1.54
75.0	1.02	1.12	1.19	1.27	1.36	2.16
90.0	1.27	1.39	1.51	1.62	1.74	---
Limits (kPa)	2.75	2.67	2.59	2.50	2.41	

Green (bold): Met both limits and goals.

To put those numbers into context, Table 10 is from NEDU's 2015 Unmanned Testing Manual[6]. The numbers in green are $\overline{P_V}$ numbers from a good performing rebreather.

On the bottom row are estimated limits for resistive effort. Based on the nature of human physiology, those limits decrease as depth and gas density increases. So, if 2.41 kPa is a $\overline{P_V}$ limit during heliox breathing at 300 fsw, it will be much less at 1000 fsw. Without a doubt, a $\overline{P_V}$ of 8.9 kPa would greatly exceed Navy limits[cc]. The consequence of that excess would be severe ventilatory limitations, likely leading to hypoventilation, CO_2 retention, hypercarbia, and potential loss of consciousness.

In other words, diving with a canister pressure drop of 58 cmH_2O, a breathing resistance of 14.7 $cmH_2O \cdot L^{-1} \cdot s$, and a so-called "work of breathing" above the maximal limit would make for a very uncomfortable and arguably dangerous dive. In fact, that is essentially guaranteed[46].

[cc] *As of 2020, NEDU's unmanned testing standards had both "goals" in terms of what is physiologically desirable, and maximum limits.*

Table 11. Converting 300 fsw experimental RE data from Table 10.

Ve	RE	Rcalc kPa	Rcalc cmH$_2$O·L^{-1}·s	ΔP kPa	ΔP cmH$_2$O
22.5	0.41	0.22	2.26	0.52	5.32
40	0.65	0.20	2.02	0.83	8.44
62.5	1.08	0.21	2.14	1.38	14.02
75	1.36	0.22	2.25	1.73	17.66
90	1.74	0.24	2.40	2.22	22.59
Max	2.41	0.33	3.32	3.07	31.29

RE= Resistive Effort from Table 10, or $\overline{P_V}$ volume-averaged pressure[dd].

Rcalc is calculated average flow resistance, and ΔP is estimated peak-to-peak mouth pressure. ΔP is of interest because that measurement is obtainable from manned testing.

From Figure 104, the estimated activity of the same bed of Sofnolime 812 would have been 2.6, compared to the Sofnolime 408 activity of 1.7. So, a 150% increase in activity comes at the cost of over 28 times increase in breathing resistance. At least, the rise in resistance calculates as being that high during a stressful dive to 1000 feet.

The Ergun equation[42], designed for chemical engineering purposes, is unlikely to match the numbers for any rebreather canister or chemical absorbent. Your results will differ. But this very successful equation shows a significant tradeoff between canister activity (duration) and flow resistance under ideal conditions and geometry.

Before each diving mission, those tradeoffs need to be seriously considered. Down deep, where breathing resistance is already high due to gas density, the *best* fine-grain absorbent may not be best for your deep dive.

[dd] *Due to the mathematical simplicity of sinusoidal waveforms from breathing machines, these conversions are relatively simple. See Appendix C for details.*

Planned Canister Durations

Suppose you were planning dives in Alaska, California, and Florida, expecting water temperatures of 40°F, 60°, and 75°F, respectively. According to the guidance from a major rebreather manufacturer, you might feel confident that your scrubber canister would effectively scrub CO_2 for 130, 180, and 220 min, respectively (Figure 111.)[ee]

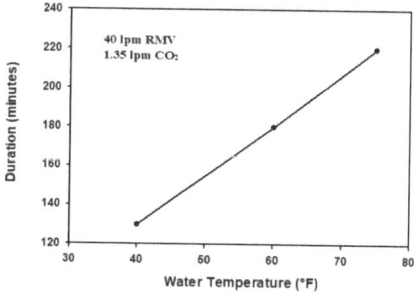

Figure 111. Canister duration recommendation from a major rebreather manufacturer.

Those manufacturer durations came from testing at a ventilation rate of 40 lpm RMV and a CO_2 injection rate of 1.35 lpm, just as the Navy Unmanned Testing Manual[6] requires for Department of Defense testing. Furthermore, the canister duration testing occurred at a simulated depth of 40 meters (130 fsw), which happens to be your planned bottom depth.

So, everything is fine. Right?

Well, not so fast. What exactly do those published durations mean? Are they the *average* canister duration?

[ee] *Figure 111 bears a remarkable similarity to Figure 69, which represented twenty runs of the SPM, five for each temperature plotted.*

The average duration means 50% of the canisters would last longer than the *average*. But it also means 50% of the canister would last a shorter time. That could easily be your fate, even if everything else corresponded to how the rebreather was tested. If you dove to the published limit, you could be inhaling far more CO_2 than you anticipated.

Since a 50% breakthrough rate is unacceptable to the U.S. Navy, NEDU publishes durations based on 95% prediction limits. That lower prediction limit is lower than the mean, or average (Figure 26). But it also means that under identical testing conditions with identical canisters, a dive that lasts up to the published limits is 97.5%[ff] likely to keep inspired CO_2 below the critical 0.5 kPa (% surface equivalent value, SEV) limit.

To illustrate these percentages, Figure 112 shows 1000 simulated canister durations distributed in a Gaussian fashion around mean temperatures[10]. In this simulation, the mean durations were curvilinearly dependent upon temperature due to an assumed granule drying effect of high temperature.

[ff] *Prediction limits are two sided, so 95% prediction limits mean 2.5% of canisters are likely to last longer than the upper prediction limit, and 2.5% of the canisters are predicted to last a shorter time than the lower prediction limit.*

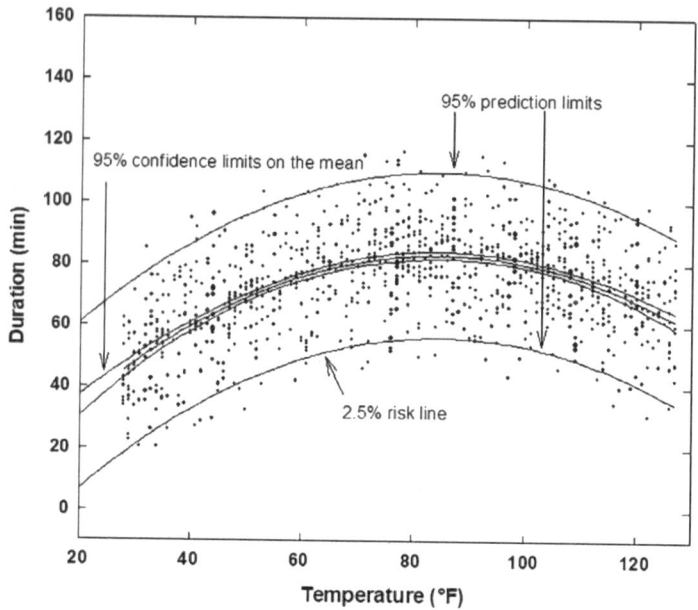

Figure 112. An illustration of mean, confidence limits on the mean, and prediction limits.

The innermost line is the best estimate for the mean duration as a function of temperature. That line is surrounded closely by the 95% confidence limits on the mean. The furthermost curved lines mark the boundary for the 95% prediction interval.

About 2.5% of the data lies above the upper prediction limit line, and about 2.5% lies below the lower prediction limit.

Of course, no testing laboratory would ever test 1000 canisters. In fact, there has always been some disagreement among engineers about how many canister duration tests are adequate.

The math is compelling when obtaining statistically-based canister durations, like prediction limits[10]. The more canisters tested at a given water temperature, the more certainty about the prediction. Furthermore, the 95%-based limit is **longer** the more units are tested.

For instance, a test of two or three canisters will yield a **shorter** canister duration than a test of five or more canisters. So, the payback for the additional cost of testing is that the users will be able to dive their canisters longer, with greater certainty than if fewer tests had been run.

Physiological Variation

So, are you out of the woods, so to speak, if you knew the 95% prediction limit for your specific rebreather? *(This author is unaware of ANY rebreather manufacturer providing that sort of measurement detail.)*

The answer, unfortunately, is no. There are sources of variation not accounted for in either Navy testing or any other known testing. Testing laboratories must make *assumptions* about relevant physiological variables, but you know what they say about *assuming*. Your dive is unlikely to be performed at the same water temperature, depth, work rate (oxygen consumption and CO_2 production), and ventilatory rate assumed by the testing laboratories.

The most recent case of uncertainty came when research divers asked the National Science Foundation's Antarctic Diving Control Board how to deal with unknown canister durations when diving under the ice in Antarctica. The Control Board's consensus was to dive to half the manufacturer's published canister durations.

Wisely, all Antarctic rebreather dives were of short duration due to the cold water. Scrubber canisters were also recharged between dives for safety, thereby preserving the canister's ability to scrub CO_2.

An excellent account of one set of experimental rebreather dives under Antarctic ice is found in reference (25) and in *InDepth*, the e-magazine of the Global Underwater Explorers[26].

If the day ever comes that under-ice divers can remain warm for more extended periods, they will likely wish to extend dive time. Before doing so, it would be wise to test canisters at 28°F, the lowest seawater temperature in polar regions. That water temperature is considerably lower than the 4°C (39°F) required for CE testing.

Figure 113. Christian McDonald and Steve Rupp prepping for an under-ice dive in McMurdo Sound, Antarctica. Photo credit: Mike Lucibella, National Science Foundation.

Figure 114. An orange dive hut was placed over a diving hole bored through Ross Sea ice. Photo by the author.

During scientific diving in polar regions, there is often an ability to store rebreathers in non-freezing temperatures between dives, such as these dive huts from a Smithsonian expedition in 2008.

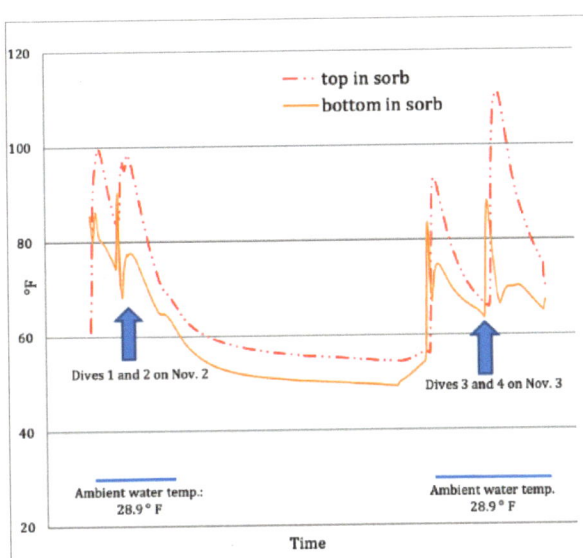

Figure 115. Canister temperature recordings in a Legacy Megalodon rebreather in Antarctica. Graphic courtesy of John Heine[25].

The minimum temperatures recorded in Figure 115 show the effects of being protected from freezing between dives.

A heated dive hut may be a luxury for divers in remote polar regions, but it does avoid the cold "false-start" issues described in Figures 87-90.

Figure 116. First Antarctic Dive Program CCR dive. McMurdo Jetty Megalodon-Poseidon-Titan rebreathers. Photo by John Heine.

Figure 117. Photo by the author.

CHAPTER 7. CFD

When university engineering departments incorporate computational fluid dynamic (CFD) models into their graduate studies, that is a sign of the growing maturation of an investigational field. The SPM modeling efforts described in this book have contributed to that progress.

The first description of the stochastic SPM[8] caught the attention of engineering students and diving hobbyists. Consequently, two engineering students pursued their post-graduate projects using experimental and CFD tools to explore CO_2 scrubber canister dynamics.

Ansys CFX 13.0

Shona Cunningham, a Medical Engineering doctoral student from the Cork Institute of Technology, generated physical and computational fluid dynamic (CFD) models of axial and radial scrubbers (Figure 118). Her CFD package was Ansys CFX 13.0.

Quoting Cunningham[50], "The body of experimental data is employed to validate the author developed transient CFD models to further the knowledge of CO_2 absorption. Static artificial neural network models are also developed and employed...The developed analyses also achieve good correlation with simulation work carried out by Clarke[8], where a direct link between the efficiency of CO_2

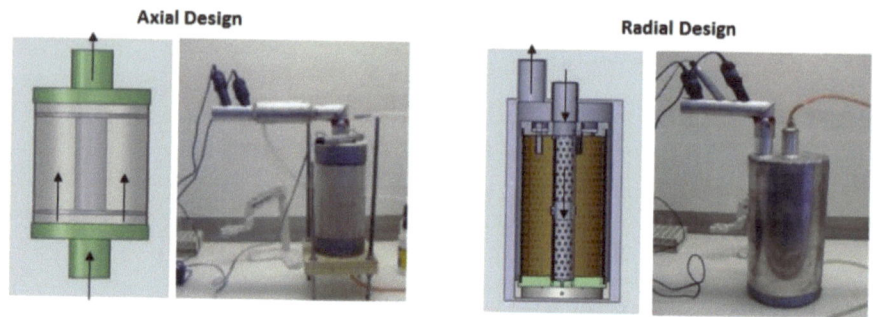

Figure 118. Axial and radial scrubber experimental set-up[50] (from Cunningham, 2013, used with permission).

absorption and cold temperatures were characterized."

In Cunningham et al.[51], after an introductory description of the NEDU SPM model, she further described her CFD work in an easily accessible paper. (At the time, Cunningham was unaware of the NEDU work on radial canisters.)

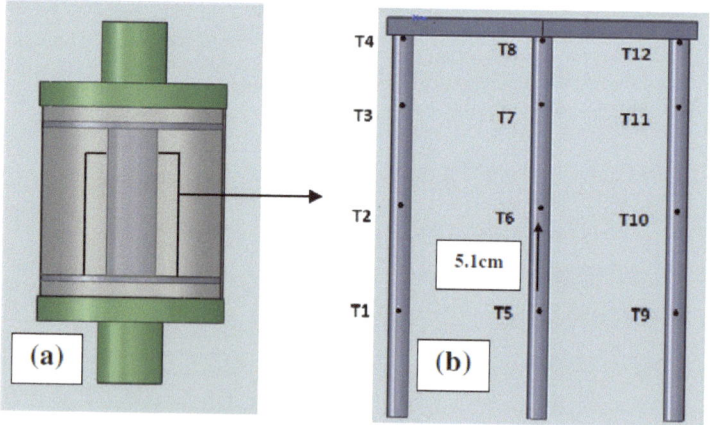

Figure 119. Test rig set-up. (a) Axial scrubber and (b) Schematic of thermocouple positions in the axial scrubber[50] (Used with permission).

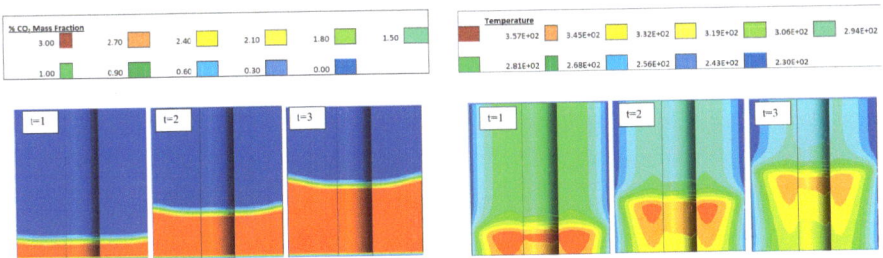

Figure 120. Cunningham's Transient Computational Fluid Dynamic model prediction of CO_2 absorption and temperature (K) profiles for an axial scrubber.[51] See the reference for details.

The validation of Cunningham's methodology (Figures 118 and 119) with U.S. Navy work[52] is illustrated in Figure 121. As expected, canister duration varied with the inverse of the CO_2 injection rate.

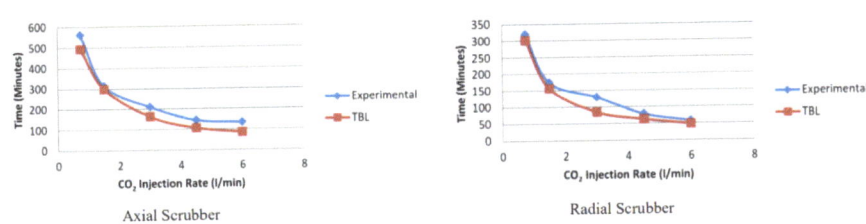

Figure 121. Cunningham[50] compares experimentally determined CO_2 injection rates with theoretical bed-life (TBL) calculations from Nuckols et al. (1996)[52].

FlexPDE

In 2001, Joerg Hess was a Master's Engineering student from Rheinisch-Westfälische Technische Hochschule (RWTH), Aachen. Germany. He had previously introduced the NEDU SPM to his Mechanical Engineering department.

For his Master's project, Hess used FlexPDE, a finite element model builder for partial differential equations. The result was a "Rebreather-Hypercapnia, (RCap)" simulation (Figure 122) for his thesis on "Development and Evaluation of a Carbon Dioxide Buildup Analyze System for Use in Closed and Semi-Closed Circuit Rebreather Diving."[53]

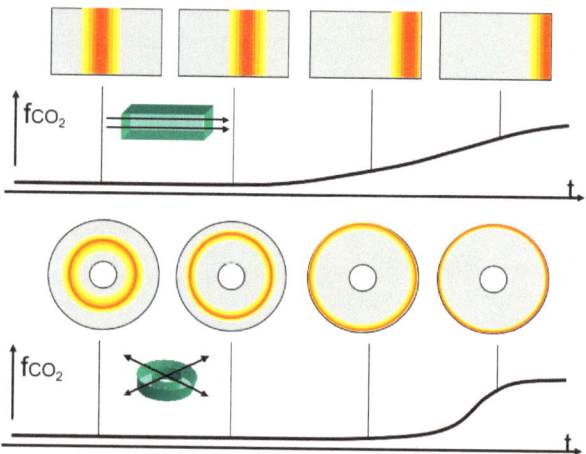

Figure 122. Heat, flow, and fraction of CO_2 in the canister eluent of axial and radial canisters.[53]

Hess's model suggests that radial canisters "break through" later than axial canisters, but once breakthrough has started, it does so more vigorously than axial canisters, all else being equal. To date, that prediction has not been confirmed.

CFD Contribution to the U.S. Navy

Inspired by the stochastic modeling work at NEDU, engineering academia applied their own deterministic tools to the problem of CO_2 absorption by rebreather scrubber canisters. By using independent experimental arrangements and protocols, the CFD work supports the NEDU mathematical modeling and stochastic SPM described herein.

The results are self-consistent, despite the widely differing analytical approaches. That consistency indicates that the U.S. Navy and rebreather divers now have a far more complete understanding of the stochastic nature of the thermo-chemical kinetic processes within rebreather canisters than before this work began. That understanding lends credence to the current statistical approach used by the U.S. Navy to evaluate the results of scrubber canister testing and for deriving diving limits for use by the Fleet.

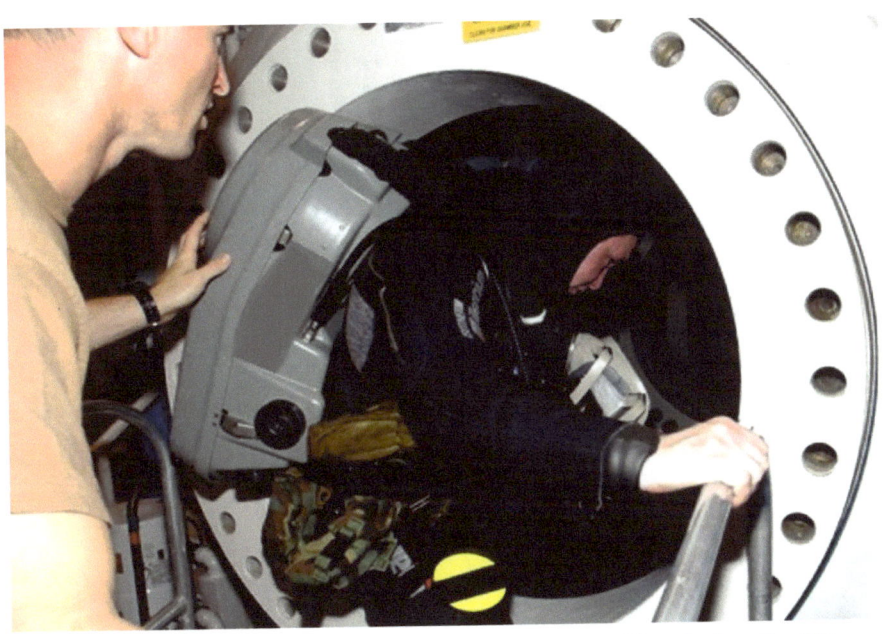

Figure 123. A Navy diver entering NEDU's Ocean Simulation Facility for an experimental dive with a MK 16 rebreather.

Figure 124. McMurdo Station from the Ross Sea. Photo by the author.

Figure 125. Boring a diving hole in the Ross Ice Shelf, Antarctica. Photo by the author.

Chapter 8. Conclusions

For warm canisters with warm inlet temperatures, CO_2 absorption fronts were narrowly defined in the SPM model, and canister duration was protracted. Lower temperatures, larger granules, and higher gas flow rates decreased canister duration. The breakthrough occurred almost immediately in a pre-chilled canister, followed by transient recovery of diffusely distributed absorption. That partial recovery was short-lived, leading to secondary absorption failure.

The SPM model replicates the essential features of CO_2 scrubber kinetics, including the effects of temperature, residence time, and canister packing. It duplicates the complex non-sustaining reaction front behavior in cold canisters. It provides a means for experimentally altering model parameters to search for mechanisms of canister behavior, including obeying Fourier's law of conduction.

Summary of Relevant Factors

The factors affecting canister function and duration are both physical and biological. Those factors were assessed at the molecular level in the stochastic SPM model and mathematically using the known variability of those factors. The propagation of errors was applied to the distributions of biological and physical factors and direct measurements made on rebreathers by the Test and Evaluation Department of the Navy Experimental Diving Unit.

Those factors include the thermodynamic properties of diluent gas mixtures such as helium and nitrogen (discussed elsewhere), the size and number of sodalime granules, and granule and bed porosity. The flow rate through the canister (inversely proportional to residence time, t_r) is critical, as is the CO_2 concentration entering a canister.

The amount of heat released by CO_2 absorption and the acid-base neutralization process is also essential to the process. Lacking good data on those numbers, we assumed values that seem to produce temperature changes like what has been found experimentally.

Of further influence are the boundary conditions for the temperature at the canister wall, which are affected by the presence or absence of insulation, water temperature, and canister design.

Physiological factors which vary from diver to diver, or day to day, are the mass of absorbent in the canister (a function of packing efficiency by divers), oxygen consumption ($\dot{V}_{O_2}$) during the dive, CO_2 production ($\dot{V}_{CO_2}$) as derived from $\dot{V}_{O_2}$, the respiratory exchange ratio, R, and the ventilation equivalent (or K) for O_2 and CO_2.

Chemical absorption processes are every bit as stochastic as deterministic. For a comprehensive verification and explanation of that statement, you need only consult Érdi and Tóth (1989.) As quoted in their Preface's second paragraph[54], "Both deterministic and stochastic models can be defined to describe the kinetics of chemical reactions *macroscopically*. (Microscopic models are out of the scope of this book)."

Well, if nothing else, the SPM shows that the randomness inherent in both physics and human physiology influences the lifetime of a CO_2 scrubber canister. Based on the illustrations in this monograph, it seems safe to say that randomness can be seen and understood on a microscopic, even molecular, level. Indeed, those familiar with statistical mechanics will know that probability and statistics are the tools used to quantify the randomness in such physical-chemical processes.

I suspect that if this work attracts the attention of those fluent in statistical mechanics and the kinetics of chemical reactions, compared to what I have offered, a much better explanation will be forthcoming.

Until theory replaces experimentation, unmanned testing laboratories will continue constraining their uncertainty by controlling many variables. Thus, they cannot gain a complete picture of what diving with rebreathers is all about.

As a result of this work, where we have explored wide-ranging environmental and physiological conditions, we come to the following actionable conclusions regarding the <u>unmanned testing of rebreathers.</u>

1. Oxygen consumption and CO_2 production should be no higher than is physiologically maintainable by an athlete performing a long-duration swim carrying mission equipment. The scrubber canister procedures used by the U.S. Navy are based on that premise.

2. To limit the effect of experimental variability, the number of replicated tests should be high enough to reduce uncertainty in measured canister durations. Under usual circumstances, NEDU requires a minimum of five replicates for each environmental condition.

3. To minimize variability in the packing of loose fill granular absorbent in a testing laboratory, ideally, only one well-practiced individual should be used for canister packing. As an active rebreather diver packing your gear, load it with the same care you would if you were packing your own parachute. Your life might depend on it.

4. As shown in references 2, 3, and 6, NEDU uses multiple water temperatures for its canister duration measurements. However, the water temperature on all real-world dives is seldom measured at the average depth of a dive. Therefore, actual canister durations may not match expectations from testing.

5. The use of 95% prediction limits[6,10] provides a reasonably safe level of conservatism for canister duration measurements under the conditions measured.

6. No statistical analysis can match the precision afforded by a gauge of CO_2 scrubber performance. Temperature-based meters provide information on the "burn" rate and may provide early warning of scrubber failure.

7. The Achilles' heel of temperature-based gauges may be the variables discussed in this book, namely, the variety of diluent inert gas species (nitrox, heliox, trimix) and gas-related changes in thermal conductivity and heat capacity. Other variable factors are diver ventilatory rate, water temperature, canister insulation, and absorbent properties (small grain versus large grain absorbent).

8. The starting temperature of the absorbent bed is critically important. Unpublished field data and both experimental and theoretical models reveal that in cold-soaked canisters, an initial pre-breathe to warm a thoroughly chilled canister can end in absorption failure after an initial burst of CO_2 absorption activity.

9. The SPM illustrates what we all suspect: insulating a scrubber canister pays off with longer canister durations in cold water. How to insulate is best left to the rebreather designer. A neoprene sleeve that works well at shallow depths on a 100% oxygen rebreather may provide little benefit to a mixed gas rebreather at depths of 100 feet or deeper.

10. Manufacturer-published canister durations are often (if not always) lacking critical information to help the diver judge the safety of a planned dive. As a rebreather owner and diver, consider

pressing the manufacturer for details missing from your User's Manual. Beyond that, it would be wise to assume that a real dive may entail far different conditions than are found in a testing laboratory. Furthermore, since divers are not mechanical breathing machines built to specification, the testing agency's assumptions about diver physiology may not match your physiology on any given dive day. So, divers beware.

11. Because of unavoidable uncertainties among laboratory testing, diving conditions, and diver physiology, an accurate sensor of both canister burn rate (heat-based sensors) and CO_2 concentration in the canister effluent would be the best insurance against an unexpectedly large load of inhaled CO_2. Some technology is currently available, but much better technologies are needed.

12. The Ergun equation provides a convenient way to estimate the effect of granular sodalime on breathing resistance. However, when comparing the calculated numbers against a physiological resistance limit or against NEDU goals and limits for resistive effort (volume-averaged pressure), the breathing resistance of mouth-pieces, valves, and hoses is not included.

13. Don't forget ALL the sources of breathing resistance in your rebreather. Good engineering design keeps non-canister sources of flow resistance to a minimum. But when a diver is deep and working hard, the resistances within a rebreather's gas flow path can add significantly to total respiratory work.

14. Don't forget ALL the sources of variability within your rebreather and your body. While engineers are familiar with the propagation of error[55], non-engineers are probably not. However, it applies to everything you do in a stressful environment, from deep breath-hold diving to deep diving on CCR. Being "above average" is not always a good thing.

Figure 126. Every rebreather component visible in this photo adds to breathing resistance. (Stock photo.)

15. Of all the sources of "biological" variability, the two a diver can most easily control are oxygen consumption and scrubber absorbent mass. Being physically fit, proficient in your diving, and relaxed with a well-maintained and properly functioning rebreather are the best ways to control oxygen consumption. This pays dividends in both breathing resistance and canister duration.

16. As was shown in Figure 107, and as explained in Appendix C, the most useful and easy to interpret measurements from both manned and unmanned dives with UBA, is peak to peak mouth pressure (ΔP). That is especially true with rebreathers, which are mechanically a relatively simple assemblage of resistive and elastic components.

REFERENCES

1. **Lillo RS, Ruby R, Gummin DD, Porter WR, Caldwell JM.** Interim contaminant limits and testing procedures for US Navy fleet soda lime. NMRI TR 95-02, Naval Medical Research Inst, 1995.

2. **Middleton JR, Thalmann ED.** Standardized NEDU unmanned UBA test procedures and performance goals, NEDU TR 03-81, Navy Experimental Diving Unit, 1981.

3. *U.S. Navy Unmanned Test Methods and Performance Goals for Underwater Breathing Apparatus*, NEDU Technical Manual 01-94, Panama City, FL, Navy Experimental Diving Unit, June 1994.

4. NATO Standardization Agreement (STANAG) 1410, edition 1, *Standard Unmanned Test Procedures for Underwater Breathing Apparatus.* 1997, updated in 2006 and 2011.

5. NATO Standardization Agreement (STANAG) 1411, edition 1, *Standard to Quantify the Characteristics of Carbon Dioxide (CO_2) Absorbent Material for Diving, Submarine and Marine Applications.* 1997, 2000, updated in 2005 and 2013.

6. *U.S. Navy Unmanned Test Methods and Performance Limits for Underwater Breathing Apparatus.* NEDU Technical Manual 15-01, Panama City, FL, Navy Experimental Diving Unit, June 2015.

7. **Clarke JR.** Contrasts between Semi-closed Circuit Underwater Breathing Apparatus of the '70s and the '90s. In: *Proceedings of Ocean Community Conference '98.* Washington D.C.: Marine Technology Society, 1998, p. 976-982.

8. **Clarke JR**. Computer modeling of the kinetics of CO_2 absorption in rebreather scrubber canisters. In: *MTS/IEEE Oceans 2001. An Ocean Odyssey. Conference Proceedings (IEEE Cat. No.01CH37295)*. 2001, p. 1738–1744.

9. *Sodasorb Manual of CO_2 Absorption*. W.R. Grace and Co., Cambridge, Mass, 1993.

10. **Clarke JR**. Statistically based CO_2 canister duration limits for closed-circuit underwater breathing apparatus Navy Experimental Diving Unit Technical Report 2-99, 1999.

11. **Clarke JR**. The analysis of sodalime granule size distributions. Navy Experimental Diving Unit Technical Report 02-08, 2002.

12. **Knafelc ME.** Oxygen consumption rate of operational underwater swimmers. Navy Experimental Diving Unit Technical Report 1-89, Navy Experimental Diving Unit, 1989.

13. **Clarke JR**. Underwater Breathing Apparatus. In: *The Lung at Depth*, edited by Lundgren CEG, Miller J. In series, *Lung Biology in Health and Disease*, edited by Enfant C. New York: Marcel Dekker, 1999, p. 429-527.

14. **Lin MJ, Jaeger MJ**. CO_2 binding by Baralyme in three different carrier gases. *Undersea Hyperb Med* 21: 329–340, 1994.

15. **Goodman M**. A study of carbon dioxide elimination from SCUBA, with standard and modified canisters of the U.S. Navy closed-circuit oxygen rig. Naval Experimental Diving Unit Research Report 1-64, 1964.

16. **Nuckols ML**, **Purer A**, **Deason GA**. *Design Guidelines for Carbon Dioxide Scrubbers*. NCSC TechMan 4110-1-83A. Panama City, FL: Naval Coastal Systems Center, 1985.

17. **Radić S**, **Denoble P**, **Gošović S**, **Živković M**. Ventilatory parameters influences on efficiency of CO_2 scrubbing. In: *Underwater and Hyperbaric Physiology IX. Proceedings of the Ninth International Symposium on Underwater and Hyperbaric Physiology. Bethesda, MD: Undersea and Hyperbaric Medical Society*. 1987, p. 547–554.

18. *Test Methods for Sodalime Carbon Dioxide Absorbents*, NEDU Technical Manual 02-01, Navy Experimental Diving Unit, 2002.

19. **Dwyer JV**, **Pilmanis AA**. Physiological Studies of Divers Working at Depths to 99 fsw in the Open Sea. In: *Proceedings of the Sixth Symposium on Underwater Physiology*, edited by Shilling CW, Beckett MW, San Diego: Federation of American Societies for Experimental Biology Bethesda, 1978, p. 167-178.

20. **Morrison JB**, **Butt WS**, **Florio JT**, **Mayo IC**. Effects of increased O2-N2 pressure and breathing apparatus on respiratory function. *Undersea Biomed Res* 3: 217–234, 1976.

21. **Morgan KR**, **Fennewald M**, Unmanned testing of Fullerton Sherwood SIVA VSW underwater breathing apparatus (UBA) for very Shallow Water (VSW) mine Counter Measure (MCM) mission. Navy Experimental Diving Unit Technical Report 06-99, 1999. https://apps.dtic.mil/sti/citations/ADA371173 [12 Oct. 2022].

22. **Meduna LJ**. *Carbon dioxide therapy: A neurophysiological treatment of nervous disorders*. Springfield: Charles C Thomas Publisher, 1950.

23. **Permentier K, Vercammen S, Soetaert S, Schellemans C**. Carbon dioxide poisoning: a literature review of an often forgotten cause of intoxication in the emergency department. *Int J Emerg Med* 10: 14, 2017.

24. **Wolfram S**. *A New Kind of Science*. Wolfram Media, 2002.

25. **Heine J, Bozanic J**. 2018. Evaluation of Closed Circuit Rebreathers for the National Science Foundation, US Antarctic Scientific Diving Program Diving for Science 2018: Proceedings of the AAUS 37th Scientific Symposium. 40-58.

26. **Menduno M**, *Bringing Rebreathers to Antarctica*, InDepth, Global Underwater Explorers. vol. 1.3, Feb 2019.

27. **Kerem D, Ariel A, Eilender E, Melamed Y**. Ventilatory response to CO_2 elevation and submerged exercise at 1 ATA in novice divers. In: *Proceedings of the Eighth Symposium on Underwater Physiology*, edited by Bachrach AG, Matzen MA. Bethesda: Undersea Medical Society, 1984, p. 493-501.

28. **Dunworth SA, Natoli MJ, Cherry AD, Peacher DF, Potter JF, Wester TE, Freiberger JJ, Moon RE**. Hypercapnia in diving: a review of CO_2 retention in submersed exercise at depth. Undersea Hyperb Med. 2017 May-Jun;44(3):191-209.

29. **Kerem D, Daskalovic YI, Arieli R, Shupak A.** CO_2 retention during hyperbaric exercise while breathing 40/60 nitrox. *Undersea Hyperb Med* 22: 339–346, 1995.

30. **Gao J, Tian G, Sorniotti A, Karci AE, Di Palo R.** Review of thermal management of catalytic converters to decrease engine emissions during cold start and warm up. *Appl Therm Eng* 147: 177–187, 2019.

31. **McAllister S, Finney M, Cohen J.** Critical mass flux for flaming ignition of wood as a function of external radiant heat flux and moisture content [Online]. https://www.fs.usda.gov/treesearch/pubs/40243.

32. **Silvanius M, Mitchell SJ, Pollock NW.** The performance of "temperature stick" carbon dioxide absorbent monitors in diving rebreathers [Online]. https://www.ncbi.nlm.nih.gov/pmc/articles/PMC6526050/.

33. **NATO.** *ADivP-03. Standard to Quantify the Characteristics of Granular Carbon Dioxide (CO_2) Absorbent Material for Diving and Hyperbaric Applications.* Sept. 15, 2016.

34. **Harvey D, Pollock NW, Gant N, Hart J, Mesley P, Mitchell SJ.** The duration of two carbon dioxide absorbents in a closed-circuit rebreather diving system. *Diving Hyperb Med* 46: 92–97, 2016.

35. **Clarke JR, Thompson LD, Godfrey RJ.** Lot variability of Sofnolime 408 carbon dioxide absorbent when tested in the cold. NEDU Technical Report 1-98, 1998.

36. **Warkander DE, Clarke JR**. Method for estimating the remaining capacity of the carbon dioxide scrubber in the Hyperbaric Oxygen Treatment Pack [Online]. NEDU Technical Report 02-13, October 2002. https://apps.dtic.mil/sti/citations/ADA442777 [21 Sep. 2022].

37. **Leva M**. Fluid flow through packed beds. Chem. Eng. 56:115-117, 1949.

38. **Purer A, Deason GA, Hammonds BH, Nuckols ML**. The effects of pressure and particle size on CO_2 absorption characteristics of High-Performance Sodasorb. Naval Coastal Systems Center Tech. Manual 349-82, 1982.

39. **Bear J**. *Dynamics of Fluids in Porous Media*. Courier Corporation, 1988.

40. **Dexter AR, Tanner DW**. Packing Densities of Mixtures of Spheres with Log-normal Size Distributions. *Nature Physical Science* 238: 31–32, 1972.

41. **Mudgil D, Barak S**. Viscosity and Sensory Acceptability of Almond Milkshake as Influenced by Sugar, Almond Paste and Corn Flour - A Response Surface Study. *Letters in Applied NanoBioScience* 10: 2483–2493, 2021.

42. **Ergun S**. Fluid flow through packed columns. *Chem Eng Prog* 48: 89–94, 1952.

43. **Silverman L, Lee G, Yancey AR, Amory L, Barney LJ, Lee RC**. Fundamental Factors in the Design of Respiratory Equipment: A Study and an Evaluation of Inspiratory and Expiratory Resistances for Protective Respiratory Equipment,

Bulletin 5339. 1945, Washington: Office of Scientific Research and Development, 1945.

44. **Mead, J**, Resistance to Breathing at Increased Ambient Pressures, In: *Proc. Symp. Underwater Physiology*, Edited by Goff L. Washington D.C.:Natl. Acad. Sci. Natl. Res. Council, pp. 112-120, 1955.

45. **Clarke JR, Jaeger MJ, Zumrick JL, O'Bryan R, Spaur WH**. Respiratory resistance from 1 to 46 ATA measured with the interrupter technique. *J Appl Physiol* 52: 549–555, 1982.

46. **Clarke JR**. Physiological event prediction in evaluations of underwater breathing apparatus [Online]. Navy Experimental Diving Unit Panama City United States. https://apps.dtic.mil/sti/citations/AD1035602 [21 Sep. 2022].

47. **Clarke JR, Survanshi S, Thalmann E, Flynn ET.** Limits for mouth pressure in underwater breathing apparatus (UBA). In: *Physiological and Human Engineering Aspects of Underwater Breathing Apparatus,* edited by Lundgren C, Warkander D. Bethesda: Undersea and Hyperbaric Medical Society, 1989.

48. **Clarke JR**. Diver tolerance to respiratory loading during wet and dry dives from 0 to 450m. In: *Lung Physiology and Divers' Breathing Apparatus,* edited by Flook V, Brubakk AO. Aberdeen, BPCC-AUD, pp. 33-40, 1992.

49. **Clarke JR**. Modeling Diver Tolerance to Breathing Apparatus, in: Weathersby PK, Gerth WA, Editors. Survival Analysis and Maximum Likelihood Techniques as Applied to Physiological Modeling. Undersea and Hyperbaric Medical Society, Kensington, MD, 2002

50. **Cunningham S**. Carbon dioxide absorption and channeling in closed circuit rebreather scrubbers. Cork Institute of Technology, Department of Mechanical, Biomedical and Manufacturing Engineering, Cork: 2013.

51. **Cunningham S, Burke A, Kelley G**. The use of mixture model theory in CFD for the chemical reaction between CO_2 and soda lime in closed circuit rebreather scrubbers. *SpringerPlus*, Vol. 2, No. 1:578-590, 2013. https://doi.org/10.1186/2193-1801-2-578

52. **Nuckols ML, Sarich AJ, Tucker WC**. *Life Support Systems Design: Diving and Hyperbaric Applications*. Pearson Learning Solutions, 1996.

53. **Hess J**. Development and evaluation of a carbon dioxide buildup analyzer system for use in closed and semi-closed circuit rebreather diving. Technical University, Aachen: 2001.

54. **Érdi P, Tóth J**. *Mathematical Models of Chemical Reactions: Theory and Applications of Deterministic and Stochastic Models*. Manchester University Press, 1989.

55. **Ku HH**. Notes on the Use of Propagation of Error Formulas, Journal of Research of the National Bureau of Standards, Vol. 70C, No. 4, 1966. [Online]. https://www.semanticscholar.org/paper/20ebd72466ffbe8f8b9f1138136d22b353592dce [11 Oct. 2022].

56. **Joye D, Clarke J, Carlson N, Flynn E** Formal descriptions of elastic loads encountered in the use of underwater breathing systems. In: *Physiological and Human Engineering Aspects of*

Underwater Breathing Apparatus. Edited by Lundgren C and Warkander D, Bethesda: Undersea and Hyperbaric Medical Society, Inc. 1989.

57. **Joye D, Clarke JR, Carlson N, Flynn E**. Formulation of elastic loading parameters for studies of closed-circuit Underwater Breathing Systems. NMRI Tech report, #89-89, 1989.

58. **Boron WF, Boulpaep EL**. *Medical Physiology: A Cellular and Molecular Approach.* Saunders/Elsevier, 2009.

59. **Otis AB, Fenn W, Rahn H**. Mechanics of breathing in man. J Appl. Physiol. 2:592-607, 1950.

60. **Otis AB**. The work of breathing. *Physiol Rev* 34: 449–458, 1954.

61. **Dubois AB**. Resistance to breathing. In: *Handbook of Physiology, Respiration Vol. 1*, edited by Wallace O. Fenn HR. Washington, D.C.: American Physiological Society, 1964, p. 451–462.

62. **Otis AB**. The work of breathing. In: *Handbook of Physiology, Respiration*, edited by Wallace O Fenn HR. Washington D.C.: American Physiological Society, 1964, p. 463–476.

Photo by Jeffrey Bozanic.

APPENDIX A: COLOR MAPPING

SPM uses green color to continuously track the number of absorbed CO_2 molecules (and thus calcium carbonate) in each of the 288,000 discrete volume elements (cells). When the SPM user downloads the current state of the sim, that download includes each cell's temperature and carbonate content.

Once the download is complete, each cell's carbonate concentration is color mapped in Slicer/Dicer on a scale of blue to tan, with tan representing the highest $CaCO_3$ content and blue the least.

Due to the three-dimensional presentation of the data, colors may be highlighted with white to portray the effect of a light source. That "lighting" accentuates the three-dimensional appearance of the data.

In Figure 81, the tabulated maximum green content in the 2 mm bed was 94 units. Mapping of that value in a blue to tan color map yields the color "Gainsboro: RGB = 221,221,209 and CMYK = 0% 0% 5% 13%. Since ink uses CMYK color palettes, the printed image has a yellowish tint.

In Figure 82, the maximum green content in the 4 mm bed was 47 units. Mapping of that value in a blue to tan color map yields the color "Orchid": RGB = 222,109,204, and CMYK = 0% 51% 8% 13%. The printed image has a reddish tint due to the high magenta level in the CMYK palette.

The RGB values reveal that there is over twice as much green (carbonate) in the 2 mm bed (G component = 221) as in the 4 mm bed (G component = 109).

Gainsboro

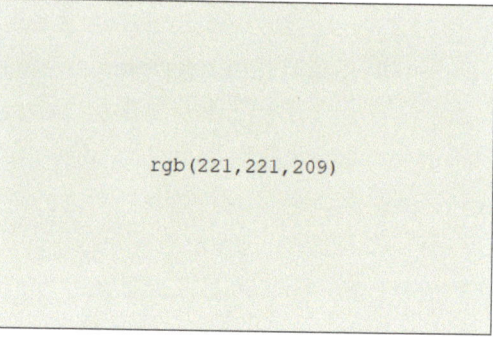

Orchid

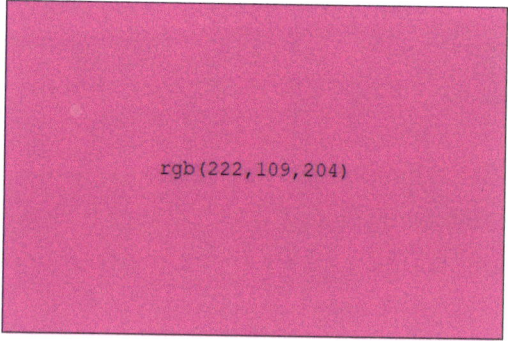

APPENDIX B: ERGUN EQUATION

Note: The following math operators and functions are formatted by PTC Mathcad Express Prime 3.1.

Small sodalime granules: 1.75 mm, σ = 0.4 mm

Ergun Equation

$$\frac{\Delta P}{L} = \frac{150 \mu V_0 (1-\varepsilon)^2}{D_p^2 \varepsilon^3} + \frac{1.75 \rho_g V_0^2}{D_p} \frac{1-\varepsilon}{\varepsilon^3}$$

- ΔP is the pressure drop
- L is the height of the bed
- Dp is the particle diameter
- ε is the porosity of the bed
- μ is the gas viscosity
- V0 is the superficial velocity (the volumetric gas flowrate divided by the cross-sectional area of the bed)
- ρg is the gas density

$cmH2O := 98.0665 \cdot Pa$ $cmH2O = 0.098 \; kPa$ definition

scrubber canister dimensions

$Lcan := 8 \cdot in$ $Dcan := 4.88 \cdot in$ $rcan := \dfrac{Dcan}{2}$ $rcan = 0.062 \; m$

$Area := \pi \cdot rcan^2$ $Area = 0.012 \; m^2$ $Area = 120.669 \; cm^2$

gas flow rate and conditions, extreme

$Vrmv := 90 \cdot \dfrac{L}{min}$ respiratory minute ventilation, volumetric gas flow rate

$V0 := \dfrac{Vrmv}{Area}$ $V0 = 0.124 \; \dfrac{m}{s}$ superficial velocity

$\rho g := 6.4 \cdot \dfrac{gm}{L}$ gas density at maximum depth

$\mu := 1.9 \cdot 10^{-4} \; poise$ helium viscosity $\mu = (1.9 \cdot 10^{-5}) \; Pa \cdot s$

absorbent bed

$Dp := 1.75 \cdot mm$ absorbent granule mean diameter (small granules)

$\xi := 0.16$ absorbent bed porosity

calculations - Ergun Equation

$$\Delta P := Lcan \cdot \left(\frac{(150 \cdot \mu \cdot V0)}{Dp^2}\right) \cdot \frac{(1-\xi)^2}{\xi^3} + \frac{Lcan \cdot (1.75 \cdot \rho g \cdot V0^2)}{Dp} \cdot \left(\frac{1-\xi}{\xi^3}\right)$$

$\Delta P = 8.17 \; kPa$ $\quad\quad\quad \Delta P = 83.315 \; cmH2O$

gas flow resistance

$R := \dfrac{\Delta P}{Vrmv}$ $\quad\quad R = 5.447 \; \dfrac{kPa \cdot s}{L}$ $\quad\quad R = 55.544 \; \dfrac{cmH2O \cdot s}{L}$

Resistive Effort, volume-averaged pressure, "work" of breathing

$Pvf := \dfrac{R \cdot (\pi^2 \cdot Vrmv)}{120}$ $\quad\quad Pvf = 0.672 \; kPa$ $\quad\quad Pvf = 6.852 \; cmH2O$

Large sodalime granules: 3.3 mm, σ = 0.2 mm

Recalculate for larger sodalime granules

$Dp := 3.3 \cdot mm$ $\quad\quad$ absorbent granule mean diameter (large granules)

$\xi := 0.32$ $\quad\quad$ porosity

$$\Delta P := Lcan \cdot \left(\frac{(150 \cdot \mu \cdot V0)}{Dp^2}\right) \cdot \frac{(1-\xi)^2}{\xi^3} + \frac{Lcan \cdot (1.75 \cdot \rho g \cdot V0^2)}{Dp} \cdot \left(\frac{1-\xi}{\xi^3}\right)$$

$\Delta P = 0.314 \; kPa$ $\quad\quad\quad \Delta P = 3.206 \; cmH2O$

gas flow resistance

$R := \dfrac{\Delta P}{Vrmv}$ $\quad\quad R = 0.21 \; \dfrac{kPa \cdot s}{L}$ $\quad\quad R = 2.138 \; \dfrac{cmH2O \cdot s}{L}$

Resistive Effort, volume-averaged pressure, "work" of breathing

$Pvf := \dfrac{R \cdot (\pi^2 \cdot Vrmv)}{120}$ $\quad\quad Pvf = 0.026 \; kPa$ $\quad\quad Pvf = 0.264 \; cmH2O$

APPENDIX C: "WORK OF BREATHING" AND FLOW RESISTANCE

An Underwater Breathing Apparatus (UBA) is composed of an assemblage of resistive, elastic, and inertial mechanical components. Of these, resistance comprises the most significant impediment to breathing.

The differential equation of motion for such a mechanical system is a second-order equation of the form:

$$\Delta P = E \cdot V + R \cdot \dot{V} + I \cdot \ddot{V} \qquad (1)$$

where ΔP is the pressure gradient across the system, V represents UBA volume, $\dot{V}$ is the time derivative of volume (flow) or dV/dt, and $\ddot{V}$ is the second derivative of volume (acceleration) or d^2V/dt^2. E is elastance, R is resistance, and I is inertance.

The ease of breathing through a UBA has long been described as the so-called "Work of Breathing," or WOB. The computation of WOB begins with the measurement of the area inside a plot of mouth pressure versus UBA volume, or:

$$W = \int P \cdot dV \qquad (2)$$

where W is the classically defined resistive Work of Breathing with units of Joules.

For our purposes, we use the equivalent form:

$$W = \int_0^T P \cdot \frac{dV}{dt} \cdot dt \qquad (3)$$

where T is the period of one breath in seconds.

When flow is sinusoidal, volume can be expressed as:

$$V = V_T \cdot \left[\sin\left(\frac{\omega \cdot t}{2}\right)\right]^2 \qquad (4)$$

where $\omega = 2\pi f_s$ and f_s is the ventilatory frequency in Hz (sec^{-1}).

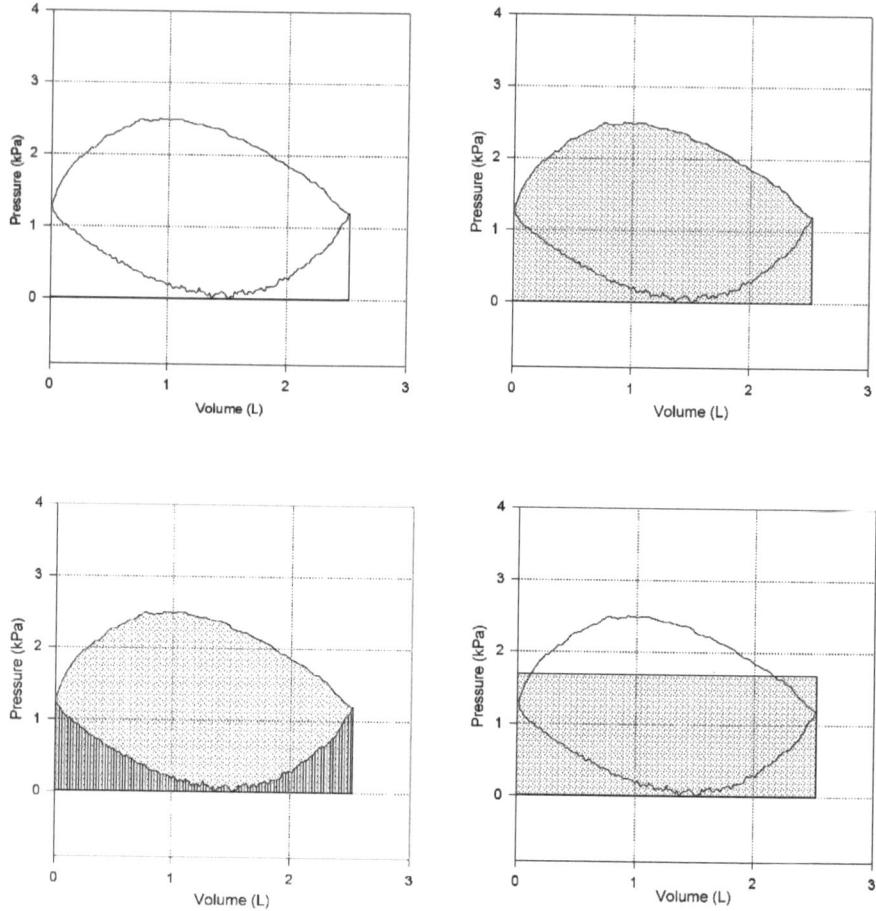

Figure C-1. Graphical interpretation of the calculation of the area inside the P-V loop.

In Figure C-1, the area in the gray rectangle (bottom right) is equal to the area (work) inside the loop. The top of the rectangle is equal to the volume-averaged pressure, about 1.8 kPa in this case.

Flow, the first derivative of volume with respect to time is,

$$\dot{V} = \frac{dV}{dt} = \frac{\omega \cdot V_T}{2} \cdot sin(\omega t) \qquad (5)$$

and acceleration, the second derivative of volume with respect to time is,

$$\ddot{V} = \frac{\omega^2}{2} \cdot V_T \cdot \cos(\omega t). \tag{6}$$

Substituting Eqns. 4-6 into Eqn. 1 and then solving Eqn. 3, it can be shown that

$$W = \frac{f_s \cdot \pi^2 \cdot V_T^2 \cdot R}{2} \tag{7}$$

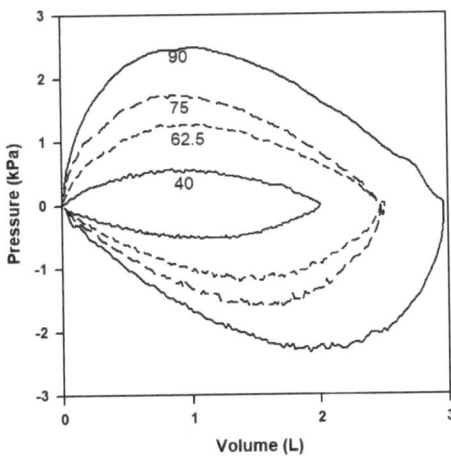

Figure C-2. Breathing machine calibration P-V loops for RMVs from 40 to 90 lpm. Each plotted loop is an ensemble average of ten loops[6].

Breathing loops are obtained at 2, 2.5, and 3-liter tidal volumes during standard U.S. Navy testing. To account for differences in tidal volumes (V_T), it has been customary in the UBA testing literature to divide W by V_T. The result is:

$$\overline{P_V} = \frac{W}{V_T} = \frac{1}{V_T} \int P \cdot dV \tag{8}$$

or mouth pressure averaged over the tidal volume, with units of kPa or J/L. This is an average value for pressure contributed by resistive impedances.

From Eqn. 7, we obtain

$$\overline{P_V} = \frac{W}{V_T} = \frac{1}{V_T} \int P \cdot dV = \frac{f_s \cdot \pi^2 \cdot V_T \cdot R}{2} \qquad (9)$$

Rearranging Eqn. 9 yields an expression for R:
$$R = \frac{2 \cdot \overline{P_V}}{f_s \cdot \pi^2 \cdot V_T} \qquad (10)$$

Expired ventilation $\dot{V}_E$, with units of L·min⁻¹, is equal to 60 times $f_s \cdot V_T$, which has units of L·sec⁻¹. Therefore, Eqn. 10 can be rewritten:
$$R = \frac{2 \cdot 60 \cdot \overline{P_V}}{\pi^2 \cdot 60 \cdot f_s \cdot V_T} \qquad (11)$$
or
$$\bar{R} = \frac{\overline{P_V}}{\pi \cdot \omega \cdot Vt/4} \qquad (12)$$

Thus, we see that average flow resistance equals average pressure divided by average ventilatory flow (Respiratory Minute Volume, RMV). We emphasize that this is an *average* resistance over the entire breath by placing a bar over R.

In engineering studies using equations such as the Ergun equation for fluid flow in packed beds (Appendix B), it is usual to solve for ΔP, the pressure drop across the packed bed. Dividing ΔP by volumetric flow yields flow resistance across the bed, R.

<u>Without Integral Calculus</u>

The sinusoidal waveform provides convenience for those not wishing to deal with integral calculus. Indeed, flow resistance and $\overline{P_V}$ can be calculated from nothing more than ΔP, the peak-to-peak pressure obtained from "breathing" a UBA.

We begin with waveforms without elastance and then provide a correction for the case of elastance: a rebreather, for example.

To find resistance directly, you divide RMS pressure by RMS flow.

RMS flow is just a peak sinusoidal flow rate from a breathing machine multiplied by the square root of two, or 0.707. Peak flow rate can either be measured with a flow meter or calculated knowing the tidal volume and breathing frequency.

Peak pressure would have to be measured by a pressure sensor, but then Prms is, once again, 0.707 times peak pressure. R is then simply Prms/Frms.

Or, you can skip that step entirely and calculate R based on volume-averaged pressure, Pva, using equation (12).

For sinusoidal flow in a purely resistive UBA, the equation for $\overline{P_V}$ is trivial.

$$\overline{P_V} = \frac{\pi \cdot \Delta P}{4} \tag{13}$$

No integral calculus is needed.

If you want to know what actual P-V work is in Joules, multiply $\overline{P_V}$ by tidal volume (Vt).

Plugging $\overline{P_V}$ from Equation (13) into Equation (12) yields average flow resistance.

$$\bar{R} = \frac{\Delta P}{\omega \cdot Vt} \tag{14}$$

ΔP is not only in the numerator of Equations (13) and (14), but it served a vital role in the assessment of the probability of eventful and uneventful dives. The ΔP measured during man-dives at both NMRI and NEDU comprised the y-axis of the meta-analysis results displayed in Figure 107. So, armed with that easy to obtain measurement from both human dives and unmanned testing of UBA, a wealth of information can be obtained.

<u>Rebreather Corrections</u>

The primary source of elastance in rebreathers is the movement of the gas/water interface in breathing bags. Without water, rebreather elastance would be minimal [55,56].

During unmanned testing, without elastance, "mouth" pressure oscillates around a zero-reference pressure (customarily taken as the hydrostatic pressure at the anatomical suprasternal notch). Rebreather elastance causes the P-V loop to be inclined, and mouth pressure rises off the zero-pressure axis.

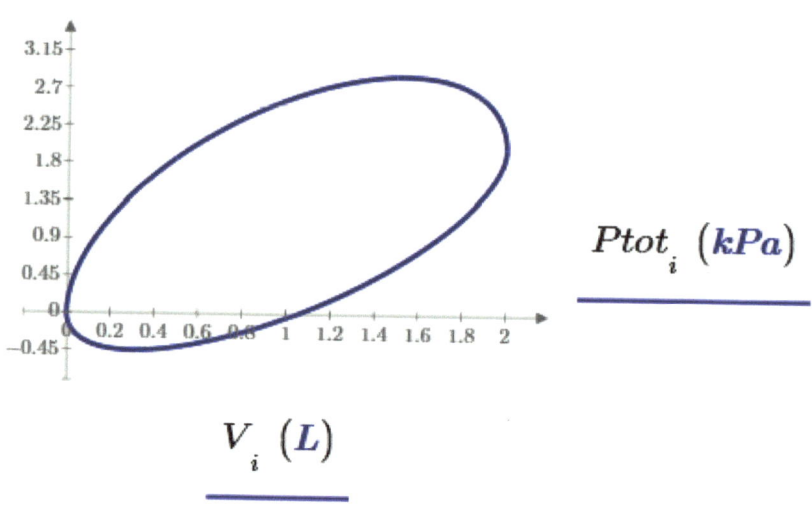

Figure C-3. Mathcad diagram of a rebreather P-V loop with elastance and unequal inspiratory and expiratory flow resistances. Vt = 2 L, $\dot{V}_E$ = 40 lpm.

When pressure waveforms are shifted upwards, the calculus-based values for $\overline{P_V}$ are not changed from the zero elastance case, but the non-calculus-based values are. They are elevated.

There is, however, an easy correction for that error. That involves taking the peak-to-peak pressure (ΔP) from the non-elastic test and dividing it by the ΔP of the elastic case. Multiplying the result of Equation (13) by the resulting ΔP ratio returns the true value of $\overline{P_V}$.

How do you find the non-elastic ΔP for a rebreather? Test the rebreather dry, without water.

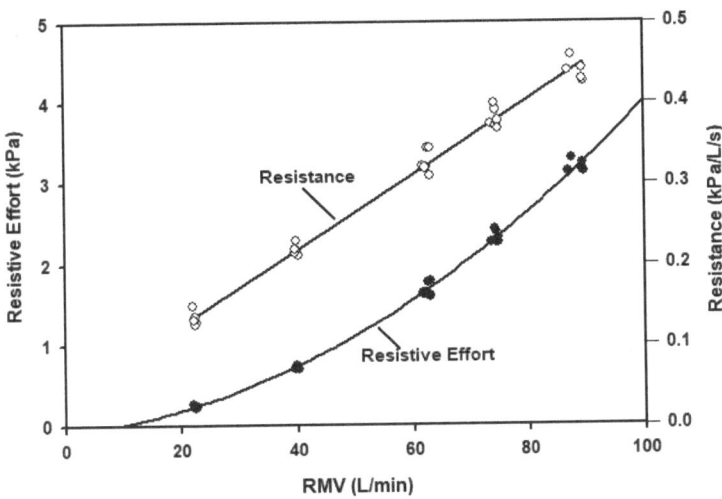

Figure C-4. Comparison of *Resistive Effort* and *Resistance* in the NEDU calibration orifice.

Remember This

If nothing else is retained from this chapter, remember the following.

In this book, *Work of Breathing* has occasionally been placed in quotation marks for a reason. It is not a measure of *work*—it is a measure of average pressure, which is why it has units of pressure, kPa or cmH$_2$O. Resistive effort (RE) or volume-averaged pressure ($\overline{P_V}$), are terms that are used interchangeably throughout NEDU Unmanned Test Manuals. They are replacements for the dimensionally incorrect term "work."

Resistive effort and *volume-averaged pressure* are not widely used outside the U.S. Navy. However, Dr. Walter F. Boron, 72nd President of the American Physiological Society, Secretary-General of the

International Union of Physiological Sciences, and co-editor of the book, *Medical Physiology*, describes inspiratory pressure as *effort*. He further divided effort into elastic and resistive components.

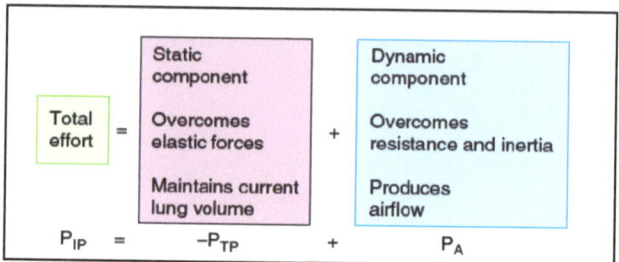

Figure C-5. *Respiratory Effort* from Medical Physiology, ed. W.F. Boron[58].

Figure C-5 shows that respiratory pressures are perceived as as *effort*, coming from both resistive and elastic sources. In Figure C-5, PIP is inspiratory pressure, PTP is transthoracic pressure, and PA is alveolar pressure.

While the phrase *work of breathing* may roll off a diver's tongue, to a physicist, physics teacher, or top-notch physiologist, that corrupted phrase is like fingernails on a chalkboard. As used by divers and some diving scientists, it is incorrect.

If you want to know about the <u>actual</u> pulmonary work of breathing, there are two sources written by preeminent respiratory physiologists in the 1950s. One paper is on *The Work of Breathing*, written by Dr. Arthur Otis and published in the Physiological Review (1954)[59,60]. That document is available on Google Scholar.

Dr. Otis happened to be my Physiology Department Chair when I was a pulmonary fellow at the University of Florida School of Medicine, Gainesville, Florida.

For a thorough grounding in the pulmonary mechanics relevant to rebreather diving, Arthur DuBois[61] has a chapter on "Resistance to Breathing," and Arthus Otis[62] has a chapter on the "Work of

Breathing" in the authoritative Handbook of Physiology, published in 1964.

Figure C-6. First seal, first served. Photo credit Jeffrey Bozanic. Titan CCR.

Figure C-7. Finally, my turn. Photo credit: Jeffrey Bozanic, Prism 2 CCR.

APPENDIX D: NUCKOLS, PURER AND DEASON

In 1983 and 1985, Nuckols, Purer and Deason published a Navy Technical Manual for the design of carbon dioxide absorbent scrubber canisters[52].

Equation (6) of that manual is the "theoretical time to consume all active absorbent." In short, "that time can be computed by dividing the absorption capacity of the chemical by the rate of CO_2 delivered to the canister." In form, it has a number of constants divided by gas stream velocity and CO_2 concentration.

$$t_{TH} = \frac{A \cdot W}{\frac{\pi}{4} \cdot D^2 \cdot V \cdot C \cdot \rho_{CO_2}}$$

A = CO_2 absorption capacity, W = absorbent mass, D = scrubber canister diameter, V = gas stream velocity, C = inlet CO_2 concentration, and ρ_{CO_2} = gas density.

For a volumetric flow rate (Q) of 40 lpm, with a scrubber diameter of 10 cm, canister cross sectional area (Acs) is 78.54 cm². V = Q/Acs, or 8.49 cm/sec. When inlet CO_2 concentration if 5%, t_{TH} is computed to be 5.32 hours.

This author derived an alternative form of the equation to yield the identical result, by substituting $\dot{V}_{CO_2}$ for C and adding A_{cs} to the numerator.

$$t_{TH} = \frac{A \cdot W \cdot A_{cs}}{\frac{\pi}{4} \cdot D^2 \cdot VCO_2 \cdot \rho_{CO_2}}$$

We chose not to use the plots of Nuckols et al, but instead plotted the alternative version of their equation, showing canister duration versus CO_2 flow rate.

Figure D-1 is an inverse first order polynomial. These types of curves are seldom published, but suit the purposes of this monograph.

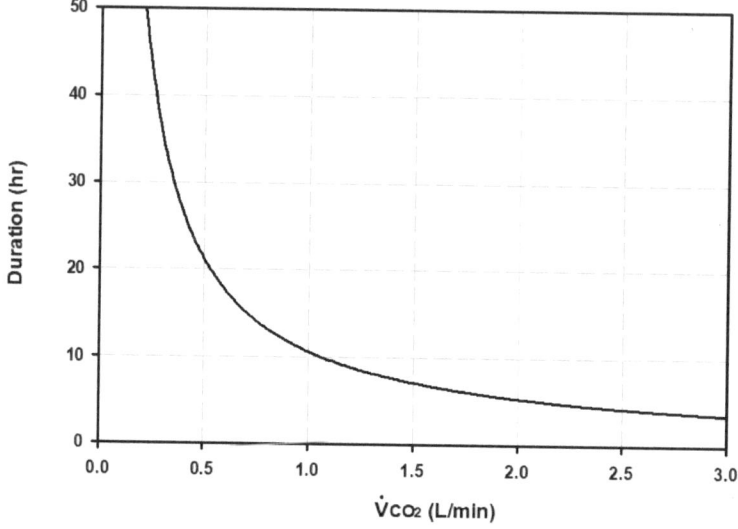

Figure D-1. Plot of theoretical canister duration.

The theoretical time to consume all absorbent at a fixed volumetric flow rate of 40 lpm and a $\dot{V}_{CO_2}$ of 2 lpm is 5.32 hours.

Notice that in the *theoretical* absorbent bed life calculation, it is possible to assume a constant ventilation rate (40 lpm, in this case) while varying $\dot{V}_{CO_2}$.

By rearranging the equations of Nuckols et al. (1983/85), we see that the hyperbolic curves shown in Appendix E are ubiquitous, despite their being largely ignored in the literature.

The reason they should not be ignored is because their region of maximum curvature seems to be associated with *clustering* or *consolidation* of CO_2 into a well-formed reaction zone.

APPENDIX E: A HYPERBOLIC CASE

The exploration of an extreme case helps us define the boundaries of the CO_2 absorption reaction, as we did in Figures 43-47. Allowing a long residence time (somewhat like a CO_2 retainer) across $\dot{V}_{CO_2}$ levels is instructive.

In Figure E-1, the water and absorbent bed temperature started at 70°F.

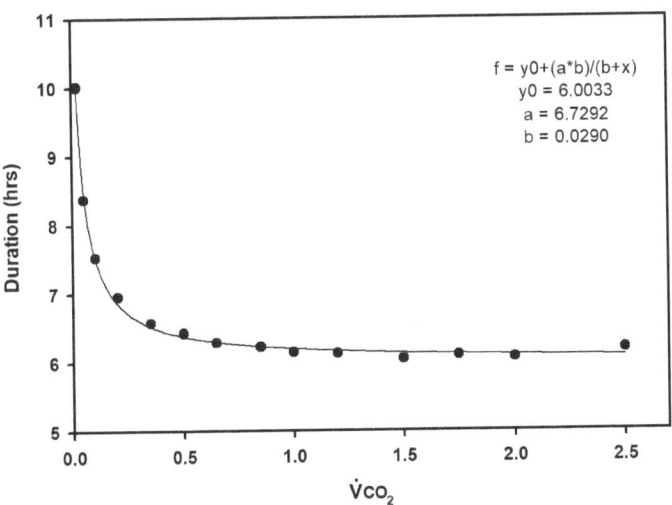

Figure E-1. A hyperbolic decay equation fit by nonlinear regression of simulated canister duration data at 70°F and Tr = 4.

$$Duration = y0 + \frac{a \cdot b}{b + \dot{V}CO_2}$$

A three-parameter model fits the data well, as evidenced by the statistics in Table E-1.

E-1

Table E-1. Statistics for the nonlinear regression of Figure E-1.

R	Rsqr	Adj Rsqr	Standard Error of Estimate
0.9985	0.9971	0.9965	0.0676

	Coefficient	Std. Error	t	P
a	6.7292	0.3432	19.6089	<0.0001
b	0.0290	0.0031	9.4154	<0.0001
y0	6.0033	0.0276	217.2358	<0.0001

There is physical significance of the three parameters in this hyperbolic equation. The parameter $y0$ is the value of duration when $\dot{V}_{CO_2}$ becomes large. It is the baseline canister duration. In turn, $y0$ is determined by absorbent bed temperature and residence time. The higher the temperature and the longer the residence time, the higher the baseline.

The parameter a determines how high is the maximum duration. When the variable x ($\dot{V}_{CO_2}$) is zero, the maximum duration is $y0 + a$.

The parameter b determines the sharpness of the transition from the rapidly declining duration to the baseline. Its effect can be best visualized in Figure E-2.

The black curve is the fit line from Figure E-1, based on simulated data at 70°F and a residence time of Tr = 4. The red and blue curves represent the effect of lowering b from 0.029 to 0.01 and 0.001, respectively.

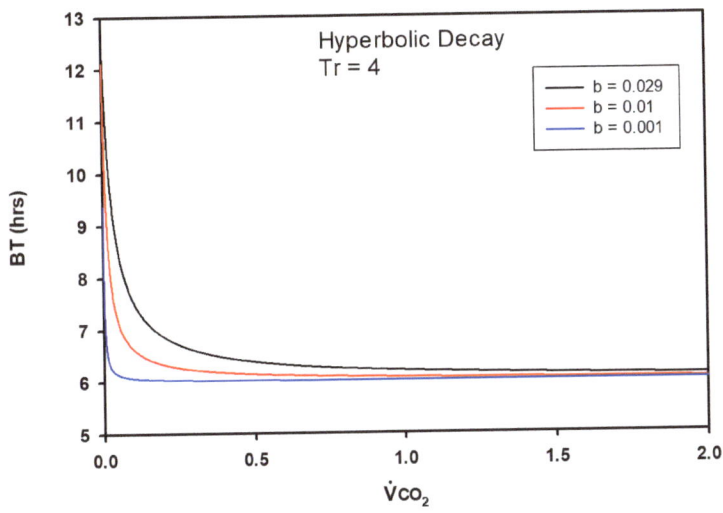

Figure E-2. The effect of decreasing b on hyperbolic curves. The y-axis is not labeled a "duration" since it is not an SPM model result.

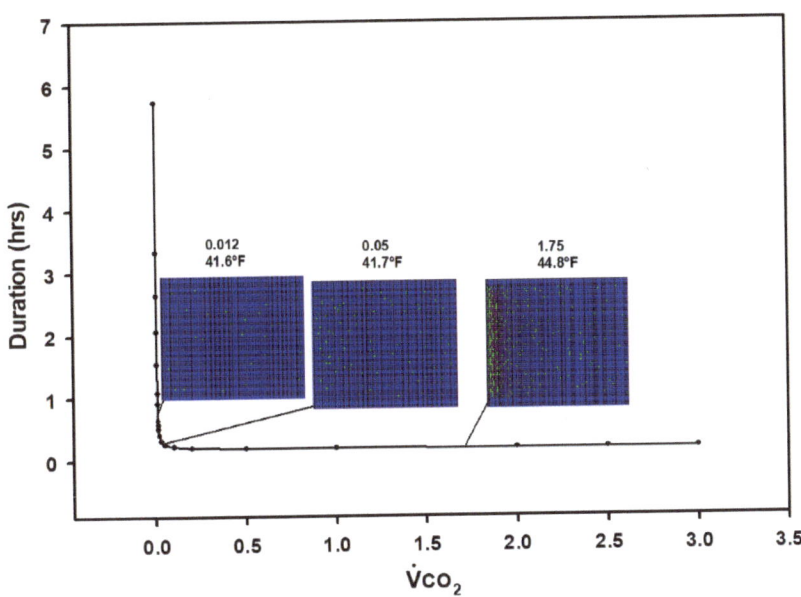

Figure E-3. Consolidation patterns and the average temperature at break through as a function of $\dot{V}_{CO_2}$. Tr = 2 at 40°F.

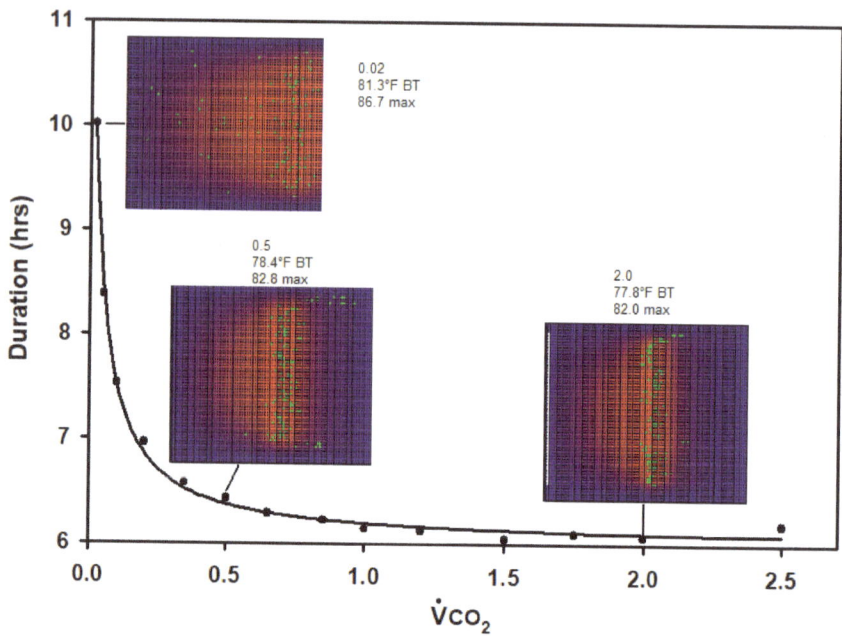

Figure E-4. Consolidation patterns and average temperature at breakthrough as a function of $\dot{V}_{CO_2}$. Tr = 4 at 70°F.

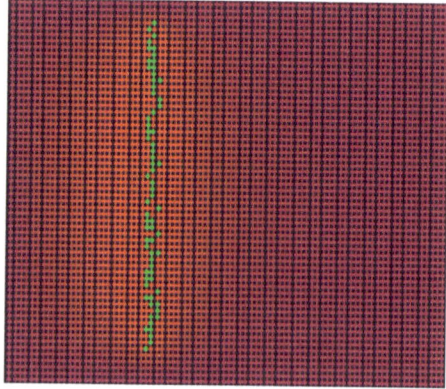

Figure E-5. Maximum consolidation at $\dot{V}_{CO_2}$ = 2.0 lpm, Tr = 10 at 100°F. 2 mm granules.

APPENDIX F: CLUSTER INDEX

Having established that reaction zone consolidation is important to the thermokinetics of CO_2 absorption reactions, we need a way to quantify that consolidation. That can be accomplished by using medial slice imagery of the "real-time" canister in the following manner.

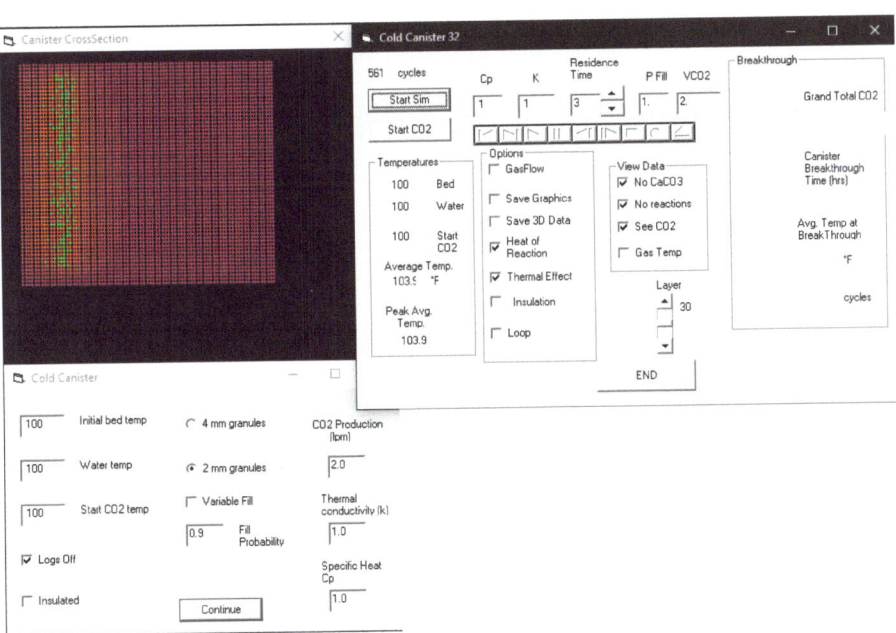

Figure F-1. A simulation run with bed and water initial temperatures of 100°F, CO_2 injection of 2.0 lpm, and a residence time (Tr) of 3.

The active reaction sites existing at computation cycle 561 are extracted from the image and digitized using the open-source image analysis software *WebPlotDigitizer* by Ankit Rohatgi.

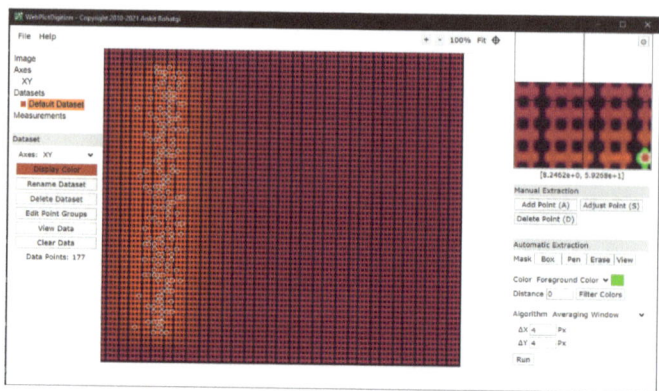

Figure F-2. Green reaction sites are isolated and digitized by open-source software.

When the resulting data file is plotted in SigmaPlot 11.0, a scattergram is obtained (Figure F-3.)

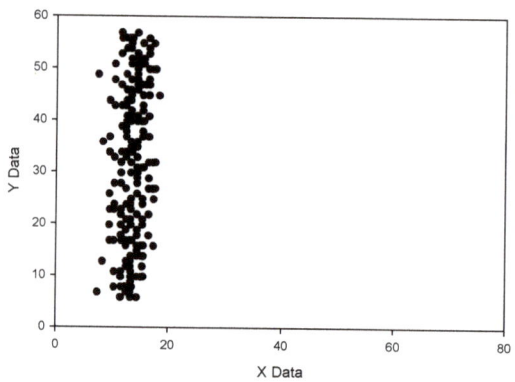

Figure F-3. Scattergram of active absorption reaction sites during one computation cycle. X data is the length of the canister, and y-data is the width of the canister and surrounding water.

Next, we determine the mean and standard deviation of the data plotted along the x-axis, column 1 (shaded row) in the following descriptive statistics table using SigmaPlot or any other statistical analysis software.

Column	Size	Missing	Mean	Std Dev	Std. Error	C.I. of Mean
Col 1	177	0	13.631	2.209	0.166	0.328
Col 2	177	0	31.562	15.137	1.138	2.245

To put that standard deviation into perspective, we examine a case for 100°F and a residence time (Tr) of 1 where there is no clustering (Figure F-4), aside from random groupings of absorption data. The measured x-data standard deviation (σ_0) was 21.552, with a mean (μ) of 40 (in the middle of the canister length.)

In this example, a Cluster Index of zero is defined relative to a standard deviation (σ_0) of 21.552.

$$Cluster\ Index = 1 - \frac{\sigma}{\sigma_0}$$

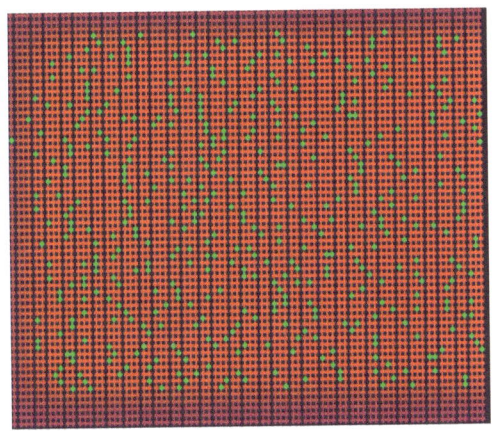

Figure F-4. 100°F, **Tr = 1**, n=447, μ = 40.073, σ = 21.552, Cluster Index = 0.000

In the following figures, Tr was 2,3,4,6 and 8, with all other initial conditions being the same as in Figure F-4.

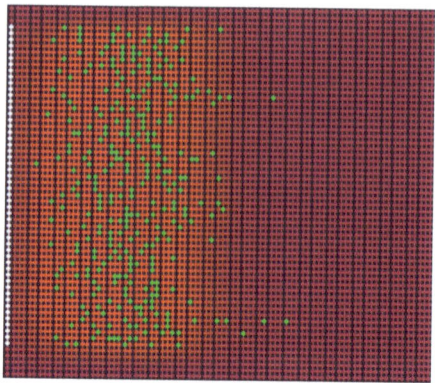

Figure F-5. 100°F, **Tr = 2**, n=315, μ = 23.703, σ = 7.992, Cluster Index = 0.629

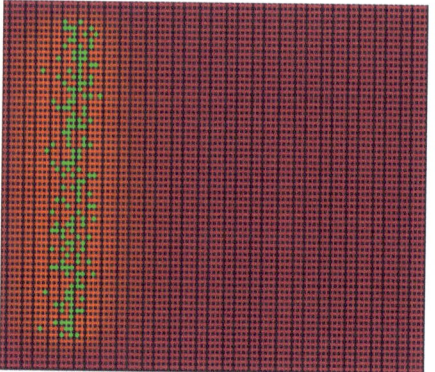

Figure F-6. 100°F, **Tr = 3**, n =177, μ = 13.631, σ = 2.209, Cluster Index = 0.898

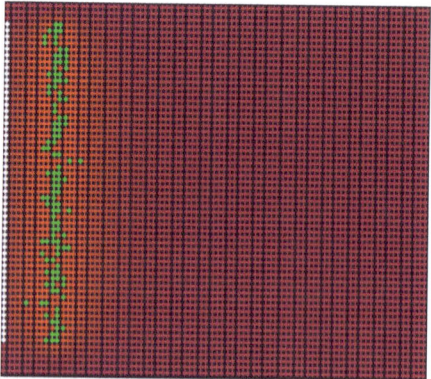

Figure F-7. 100°F, **Tr = 4**, n=141, μ = 10.105, σ = 1.245, Cluster Index = 0.942

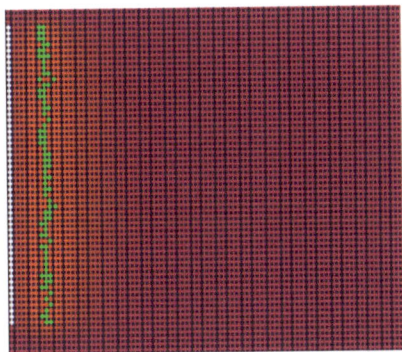

Figure F-8. 100°F, **Tr = 6**, n=114, µ = 7.965, σ = 0.683, Cluster Index = 0.968

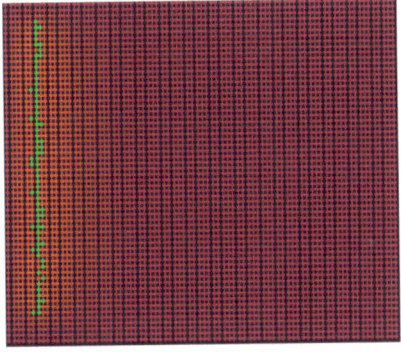

Figure F-9. 100°F, **Tr = 8**, n=72, µ = 6.707, σ = 0.547, Cluster Index = 0.975

Figure F-10. Brinacle. McMurdo Jetty and Titan CCR. Photo credit Jeffrey Bozanic.

Figure F-11. John Heine taking notes after a rebreather dive at McMurdo. Rob Robbins is looking on. Photo by: Mike Lucibella, National Science Foundation.

APPENDIX G: SUPPLEMENTARY IMAGES

This appendix contains interesting images that did not make it into the text.

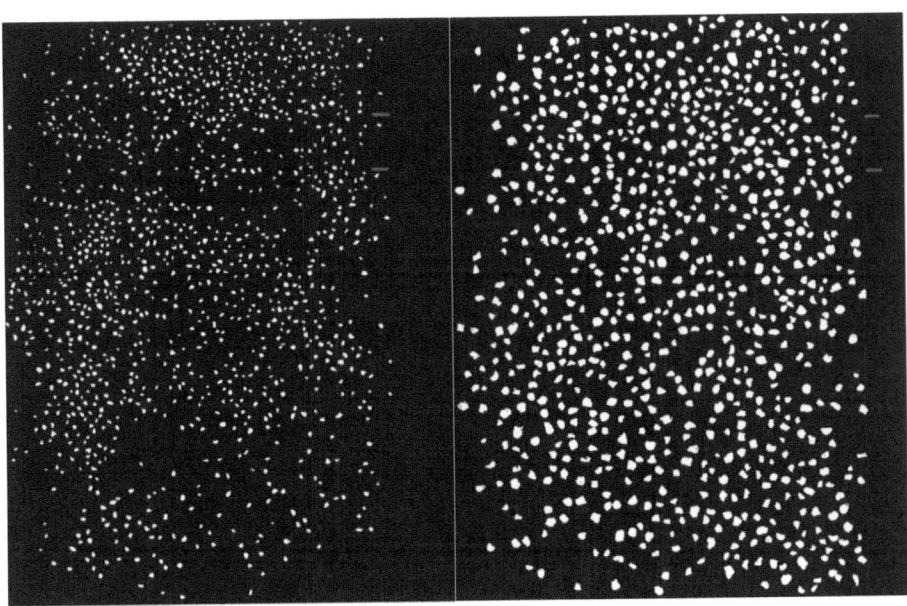

Figure G-1. Fine grain sodalime on left side and large grain sodalime on the right. Red calibration marks are 20 mm apart.

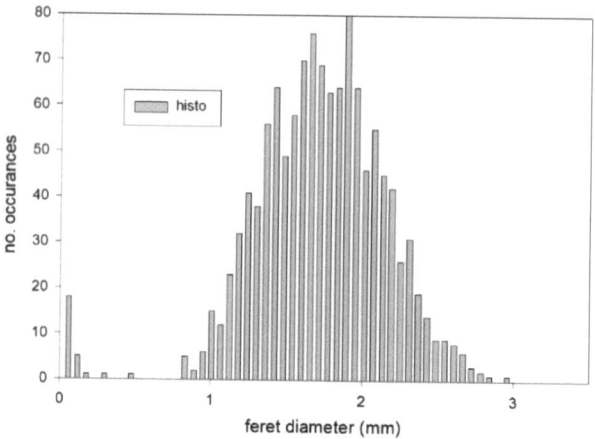

Figure G-2. Feret size for Sofnolime 812. Normal distribution.

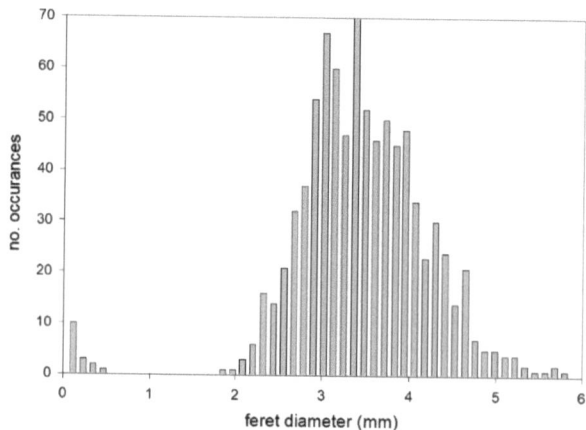

Figure G-3. Feret size distribution for Sofnolime 408. Lognormal distribution.

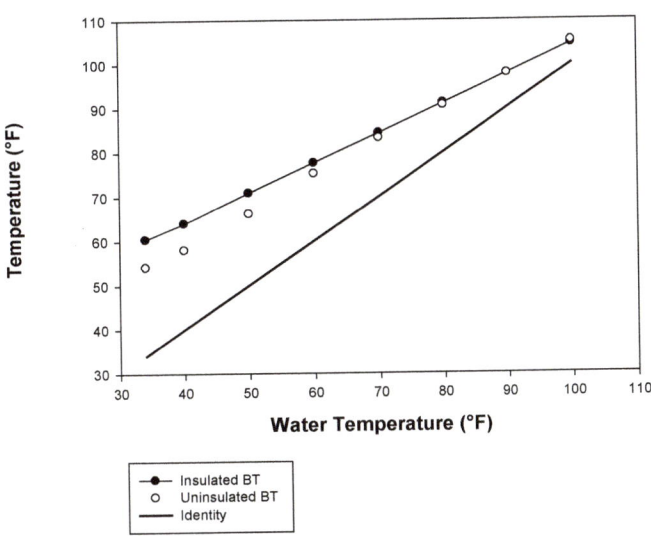

Figure G-4. Average absorbent granule temperature at canister break through (BT). The identiy line represents the canister equilibrium temperature in the absence of exothermic reactions.

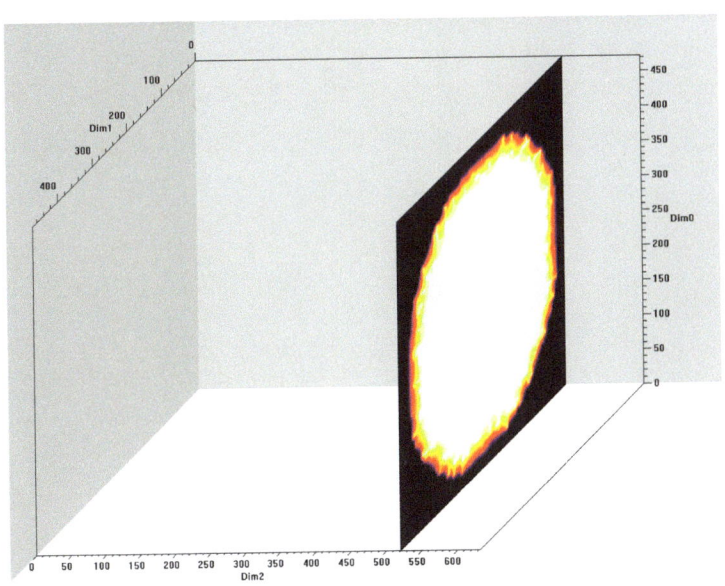

Figure G-5. The plane of maximal CO_2 absorption reactions at the moment of breakthrough in 105° F water, without insulation.

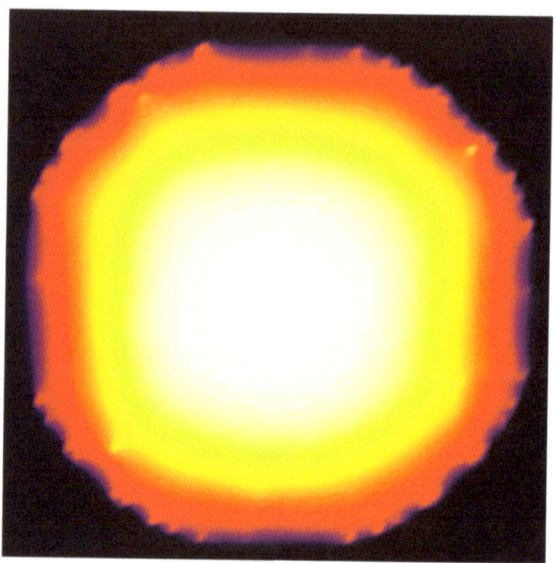

Figure G-6. Prior to reaction front arrival. Granules are heated by the convection of hot gases and the conduction of heat from upstream granules.

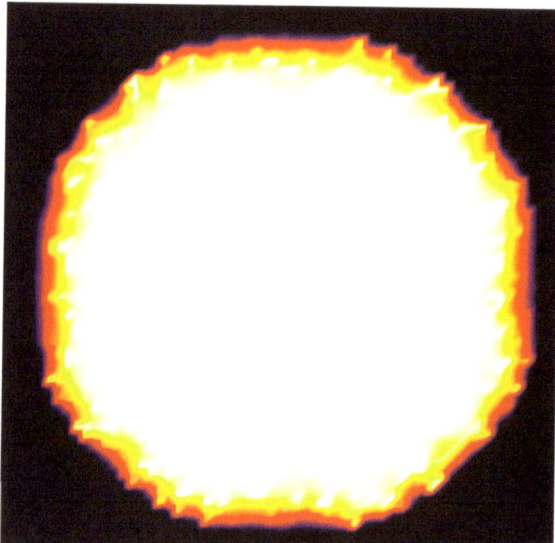

Figure G-7. The center of the reaction front near the end of the cylindrical canister in 105° F water.

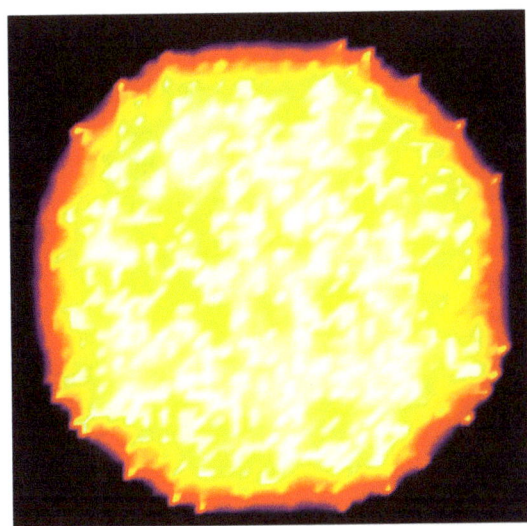

Figure G-8. Beginning cool down after passage of the reaction front.

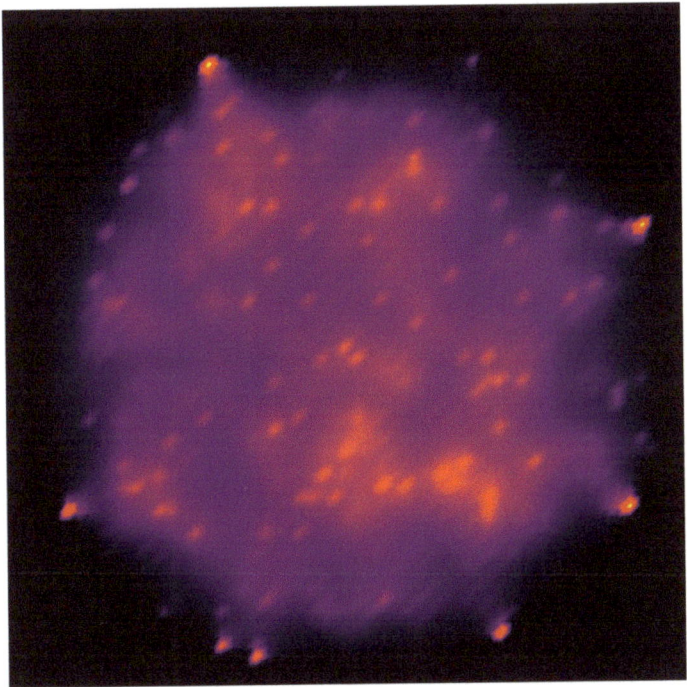

Figure G-9. Like dying embers from a fire, absorbent is cooling as the reaction front has moved past. Cylindrical canister in 105° F water.

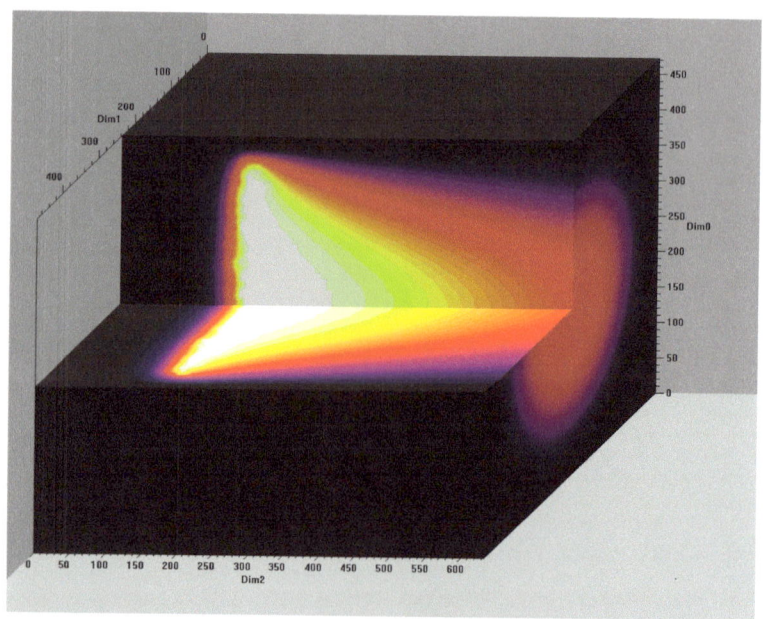

Figure G-10. Thermo-neutral at 100° F, 2.0 lpm, Tr = 4.

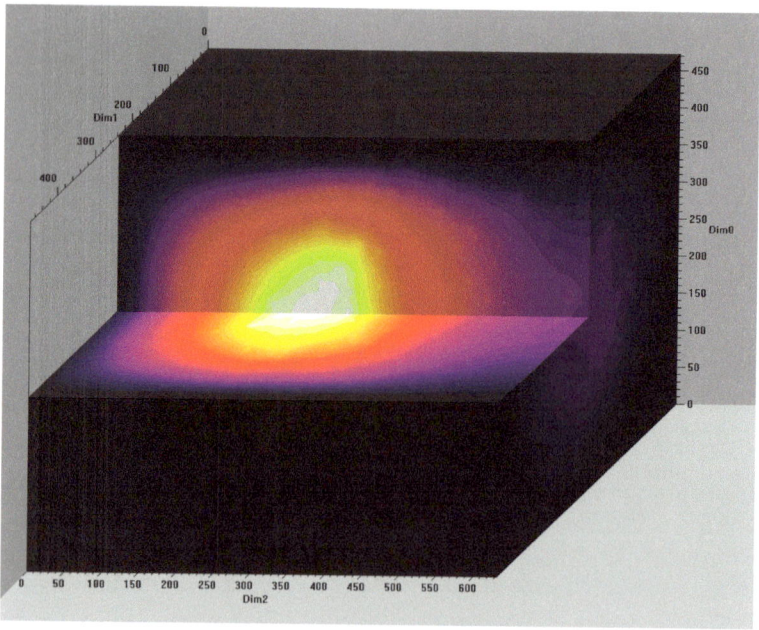

Figure G-11. Tr = 4. $\dot{V}_{CO_2} = 0.009$.

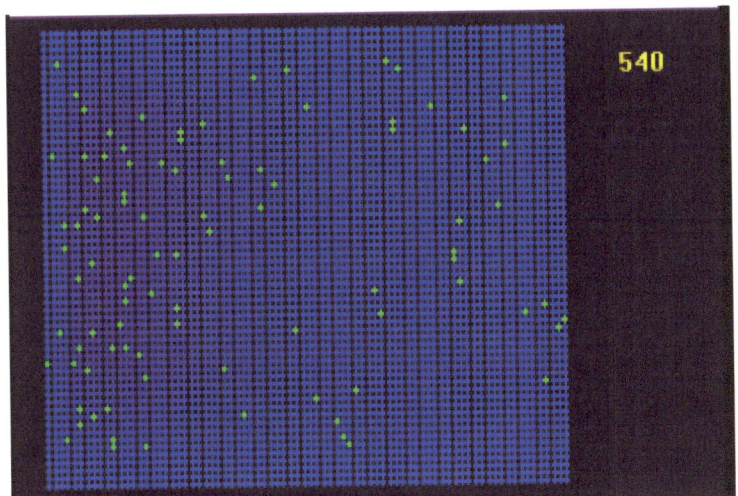

Figure G-12. $\dot{V}_{CO_2} = 0.02$, 40°, Tr3

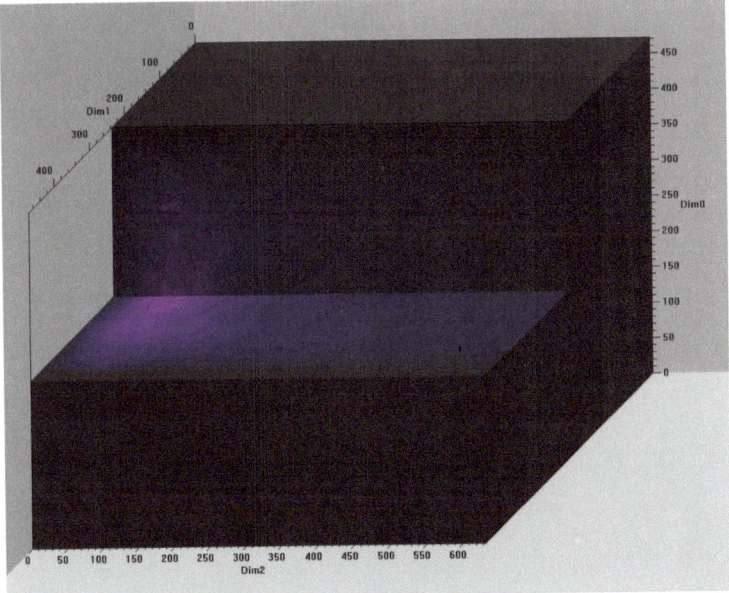

Figure G-13. Nebulosity, $\dot{V}_{CO_2} = 0.02$, 40°, Tr3

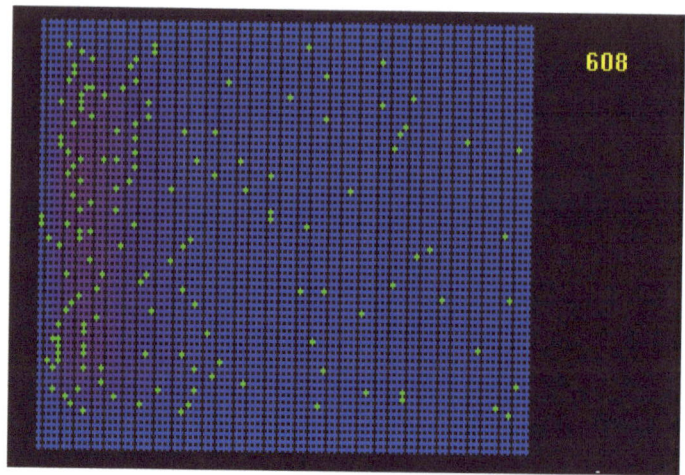

Figure G-14. $\dot{V}_{CO_2}$ = 0.05. Consolidation at Breakthrough.

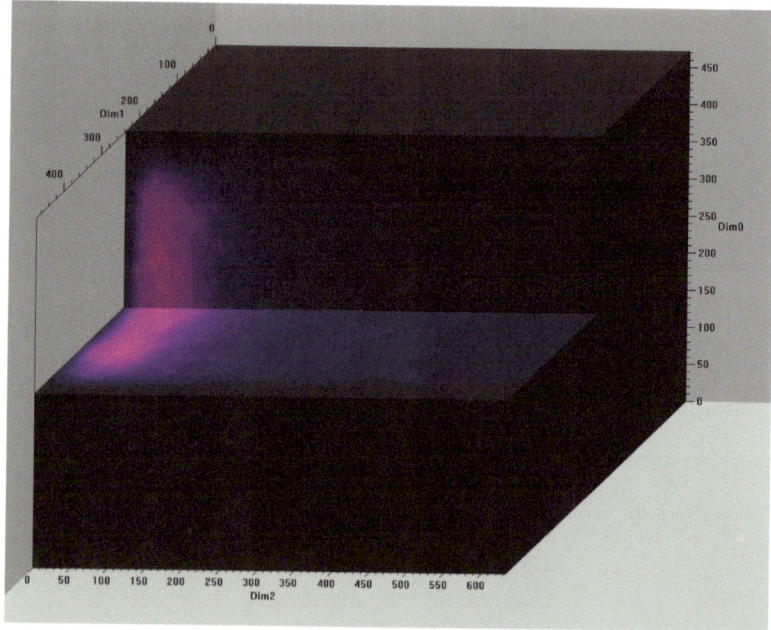

Figure G-15. $\dot{V}_{CO_2}$ = 0.05, Consolidation at Breakthrough. Temperature scale, 40°F to 130°F. Tr = 3.

INDEX

3

3D data, 70, 86

A

A.P. Diving, 74
Activation energy, 68
Activity, 74, 136, 137, 138, 151, 168
ADivP-03, 124
ADivP-05, 3
Ansys CFX, 159
Arrhenius equation, 68, 69
ATA, 25
Automatons, 115
Avogadro's number, 96
Axial, 63, 65, 67, 70, 75, 85, 89, 99, 100, 159, 160, 161, 162

B

Boron, 7, 8
Bozanic, ii, iv, v, vii, 7, 55, 56, 174, 180, 9, 10, 6
break through, 45, 162
Breakthrough, 28, 30, 37, 46, 48, 51, 52, 85, 93, 94, 95, 99, 100, 102, 121, 162
breathing resistance, 135, 136, 151

C

Canister durations, 8, 9, 10, 11, 13, 15, 16, 37, 43, 44, 46, 47, 69, 103, 104, 167, 168
Canister Insulation, 85
Canister limits, 11, 46

Carbonate, 68, 105, 108, 110
Carbonic acid, 68
catalytic converter, 82
catalytic converters, 82, 83
Cell temperature, 69, 70, 73
cellular automatons, 115
CFD, 159, 160, 162, 178
Channeling, 84, 85
Cluster Index, 3, 4, 5
CO_2 absorbency, 37
CO_2 production, 11, 15, 16, 166, 167
Coefficients of variation, 37
color mapping, 87, 107
Computational cycle, 66, 93
Conductivity, 122
Consolidation, 3, 4, 8
COV, 37, 38, 43, 51
critical mass, 110
Cylinder, 87, 90, 120

D

Dexter, 137
Diffuse, 59, 60, 61
Diffusion, 62, 63, 65
Dive Rite, 65
diving research, 143, 2
Dräger, 18
Drägersorb, 8
Duration, 8, 10, 13, 16, 29, 37, 38, 43, 45, 46, 51, 92, 93, 95, 96, 97, 98, 103, 104, 105, 161, 165, 167, 168
dusting, 124
Dwyer and Pilmanis, 51

E

elastance, 1
Enthalpy, 68
Ergun equation, 140
exothermic, 79, 82, 120, 123, 3
Exothermic, 64, 67, 68, 99, 101
Explorer, iv, 74, 132
Exponential, 29, 69

F

FlexPDE, 161, 162
Flow resistance, 137, 140, 151
Fourier's law, 123, 165
friability, 124

G

Gas constant, 68
Gaussian distribution, 11
Granules, 8, 12, 16, 28, 60, 61, 62, 63, 71, 84, 91, 95, 97, 99, 101, 120, 121, 134, 135, 136, 137, 141, 165, 166

H

Heat bubbles, 71, 73
Heat Capacity, 95, 168
Heine, v, 66, 157, 174
HYPERBOLIC, 1

I

ignition, 110
inertance, 1
Inspiration, 74

J

Joerg Hess, 161

K

Kinetics, 165
Knafelc, 172

L

Leva, 134, 135
Lillo, 1
Lucibella, v, 156

M

manned-testing, 151
Mass, 31, 36, 39, 43, 52, 58, 59, 62, 166
Mathcad, 52
McDonald, 156
McMurdo, 156
Megalodon, 133, 157
meta-analysis, 144
MK 25, 18
Monte Carlo, 52, 53
Morrison, 51

N

NATO, 171
NEDU, 2, 6, 7, 16, 17, 59, 74, 102, 104, 119, 131, 160, 161, 162, 163, 167, 168, 171, 173
Neutralization, 67, 68, 166
New Kind of Science, 58
NMRI, xviii, 6, 143
Nuckols, 161

O

Oxygen, vi, xviii, 11, 16, 18, 36, 39, 43, 44, 52, 96, 166
Oxygen consumption, 11, 16, 40, 43, 44, 52, 166

P

Packing density, 137
PHB, 71, 72
Porosity, 84, 137
Prediction intervals, 12, 43, 44, 45
Probabilistic, 62
Probability, 43, 50, 59, 62, 65, 67, 68, 69, 85, 93, 97, 98, 103, 108, 110, 167
Propagation of error, 8, 46, 47, 48, 51, 52
Propagation of Error, xviii, 13, 45
PTC Mathcad, 52
PTC Mathcad Prime, 52

R

Radial, 63, 70, 74, 100, 101, 159, 160, 162
Rate constant, 69
RCap, 162
RE, 7
Reaction, 60, 61, 62, 63, 64, 65, 68, 69, 71, 86, 90, 91, 99, 101, 105, 108, 110, 120, 121, 122, 123, 132, 165, 178
resistance, 63, 150, 151, 1, 4
respiratory effort, 8
Respiratory minute volume, 10
Respiratory quotient, 40
rEvo, 74, 132
Rgas, 31
RMV, xvi, xix, 10, 15, 16, 152, 4
Rupp, 156

S

Sentinel, 74, 132
Shona Cunningham, 159
Simulated Physical Model, 59
Slicer/Dicer, 64, 70
Sodasorb, 8, 9
Sofnolime, xvii, 12, 13, 136, 137, 141, 151
SPM, 59, 69, 70, 74, 96, 97, 98, 101, 104, 105, 120, 123, 160, 162, 163, 165, 168
SRDS, 131
STANAG, 171
STANAG 1411, 124
standard deviation, 16, 37, 97, 98, 137, 141
statistical mechanics, 167
Stephen Frink, 17
Stochastic, 59, 72, 74, 93, 99, 159, 162, 163, 165, 166
STPD, 15, 31, 96

T

TableCurve 2D, xix
TempStik, 74, 84, 111, 114
Thermal imagery, 87
Thermal map, 86

V

Variability, 8, 11, 13, 16, 37, 38, 43, 45, 46, 52, 97, 98, 165, 167
$\dot{V}_E/\dot{V}O_2$, 40
ventilatory equivalent, 35
Ventilatory equivalent, 40
Viscosity, 141
Visual Basic, 59
Visual Basic 6.0, 59
$\dot{V}O_2$, 16, 31, 40, 43, 166
volume-averaged pressure, xviii, 149, 151, 7

W

Wall Effect, 84
Warkander, 132
WebPlotDigitizer, 1

Wolfram, 57, 58

Δ

ΔP, 143, 151, 1, 4

ABOUT THE AUTHOR

A much younger John Clarke upon entering civilian Naval Service.

John R. Clarke, Ph.D. is a diving life support scientist and Fellow in Undersea and Hyperbaric Medicine (FUHM). He won the 2022 NOGI Award for Science from the Academy of Underwater Arts and Sciences. Clarke served the Navy for over 39 years, and conducted numerous research studies on dives as deep as 450 meters. In his spare time, he authored three novels about deep saturation diving.

After 27 years as the Scientific Director of the Navy Experimental Diving Unit (NEDU), he retired, then returned to Federal Service to assist the COVID Task Force. Under *Clarke Life Support Consulting*, he now consults for various defense contractors.

Clarke participated in Smithsonian/National Science Foundation (NSF) Polar diving programs in the Arctic and Antarctic. He served on the Diving Control Board for the NSF Antarctic Science Diving Program. He also chaired the Diving Control Board of FSU's Advanced Diving Program.

Before moving to Florida, he conducted diving research for twelve years at the Naval Medical Research Institute, Bethesda, MD. Earlier, he was a Parker B. Francis Foundation Fellow in Pulmonary Research in the Department of Physiology, University of Florida School of Medicine. While there, he conducted physiological research on a record-setting NEDU dive to 1500 fsw.

Before that, he was a Research Associate in Biophysics at Case Western Reserve School of Medicine in Cleveland.

His B.S. and M.S. were in Applied Biology at Georgia Tech. He earned his Doctorate in Physiology at Florida State University. Clarke was a graduate of the Navy and NOAA-sponsored Scientist-In-The-Sea program of the 1970s.

His two children and one grandchild are divers. His son-in-law is a NOAA research biologist and diver.

While a fan of all rebreathers, his preferred diving rig is a semiclosed rebreather designed for military service, the Dräger Dolphin featured on the cover of this book.

He may be contacted at john@johnclarkeonline.com.

www.ingramcontent.com/pod-product-compliance
Lightning Source LLC
Chambersburg PA
CBHW042049290426
44110CB00001B/4